JOY ADAMSON'S AFRICA

By the same author

BORN FREE
LIVING FREE
FOREVER FREE

THE SPOTTED SPHINX
PIPPA'S CHALLENGE

THE PEOPLES OF KENYA

JOY ADAMSON'S AFRICA

by Joy Adamson

Collins and Harvill Press
London

ACKNOWLEDGEMENTS

The author is grateful to the Trustees of the National Museum, Nairobi, for permission to reproduce the illustrations on pages 59—73; the Municipal Council of Mombasa for the loan of some of the paintings of coral fish reproduced between pages 40 and 48; and to Mrs Joan Waddington for permission to use the Regal Sun Bird. In addition I would like to thank Mr Robert Ardrey and his publishers, William Collins Sons & Company Limited, London, and Atheneum Publishers, New York, for permission to quote from Robert Ardrey's *African Genesis*, © 1961 by Literat S.A.

© Text and Paintings: The Trustees of the Elsa Wild Animal Appeal, 1972

© Paintings on Pages 59, 60, 61, 64 and 73: The Trustees of the National Museum of Kenya, 1972

Reprinted 1978

ISBN 0 00 262340 4

Set in 12 pt Garamond

Designed and produced by Banks and Miles, London

Made and printed by Brüder Rosenbaum, Vienna, Austria
for Collins, St James's Place and
Harvill Press, 30 A Pavillion Road, London SW 1

To all who love Kenya
and help to preserve its
natural attractions

CONTENTS

INTRODUCTION 11

1 **THE FLORA OF KENYA** 15

2 **A FEW BIRDS** 37
Regal Sunbird   Crowned Crane   Verreaux's Eagle Owl

3 **CORAL FISH** 43

4 **WONDERS OF NATURE** 58
A simulated flower   Stick Insect
Cowrie Shell   Chameleon
Agama Lizard

5 **A RECORD OF TRIBAL TRADITIONS** 54

6 **SOME ANIMALS** 66
Rock Hyrax   Leopards   Two Baby Elephants
An Impala   Arabian Oryx   A Young Buffalo

7 **LIONS AND CHEETAHS** 98
Elsa and her cubs
Pippa and her cubs

8 **A COLOBUS FAMILY** 107

SKETCHES 111

INTRODUCTION

I FIND IT DIFFICULT TO INTRODUCE MY SKETCHES WITHOUT first explaining the various interests which have led me into so many different worlds. As I grew up, I found myself torn, like so many others, between an insatiable curiosity about the bewildering world around me and an urge to discover the medium through which I could express myself and in doing so be constructive.

Though I was born in Troppau, then the capital of Austrian Silesia, and brought up in Vienna, we spent our holidays at Die Seifenmühle near Troppau, an estate belonging to my mother's family. It represented home to me and this was where my roots were.

As a child I enjoyed roaming across the fields and through the forest. Often I was accompanied on my walks by an Alsatian, but I preferred going alone for the dog chased away the wild animals I loved to watch. Best of all I liked going out with the gamekeeper and learning from him about the deer, the hares and the other creatures we saw. Shooting and hunting were part of our way of life but I liked this aspect less.

Austria is an intrinsically musical country and in our family each member either played an instrument or sang. As early as I can remember I listened to chamber music and to amateur productions of opera and went to sleep to these arias, rather than to the nursery rhymes which most children know. I could play the piano before I could read or write and I took it for granted that in one way or another I should carry on the family tradition.

When I was older, I did not wish merely to play an instrument, but decided to study for the State Diploma for pianists, which meant learning the history of music, composition, counterpoint and so on. I started the course when I was fifteen and passed the exam at seventeen. By then I had discovered that my hands were too small for me to become a concert pianist and so, since I did not want to become a teacher, I realised that I must find my fulfilment in some other field.

Many and varied were those I investigated: I studied metal-work, making ornamental bowls of brass and silver; embossing taught me to see form in relief. From this I passed on to designing posters and book-jackets, then, being interested in fashion, I did a two-year course in dressmaking for which I gained the Gremium diploma. Simultaneously I took singing lessons, which I very much enjoyed and I also helped to excavate a prehistoric site.

In the evenings I attended courses on the history of art and on drawing from life, and I also watched the restoration of Quattrocento pictures; thus I learned something about the complex craftmanship used by the old masters when mixing their pigments. Finally I took a course in shorthand and typing. All these occupations were no more than explorations through which I hoped to discover my true fulfilment.

Holidays spent playing tennis or sunbathing by the pool at Die Seifenmühle seemed a dreadful waste of time, so during one of them I went over to see a sculptor who lived nearby and who had recently completed a marble sarcophagus for our family vault. I was impressed by the design of the figures but felt more attracted to portraying the living than the dead. My first woodcarving was of a woman cuddling a hare.

Modelling in clay never gave me the same satisfaction as releasing a form concealed in stone or wood and soon I found that what satisfied me most was working in wood. This was because, by following the grain of the timber, I was able to enhance the beauty of the form; I also felt that wood was more alive and responsive than stone. Not long afterwards I became a pupil of Professor Frass in Vienna. He was remarkable for the fact that he was able to carry out all the work himself (whether he was handling marble or bronze), from the initial design to the final monument. From him I learnt the importance to an artist of really understanding his medium, for the medium – whether wood, stone, or terra cotta – is fundamental to the ultimate design. Although I enjoyed sculpting immensely there came a time when reproducing the outer form of the human body no longer satisfied me. I wanted to know its psychological motivation and physical structure. My curiosity led me into the dangerous field of psychoanalysis (of which Vienna is the cradle), and to the dissecting-room in the Anatomical Institute.

Friends had warned me that I might faint when confronted by a corpse but in fact I was so fascinated by the miraculous interaction of muscles, tendons and bones that I determined to study medicine.

Before I could do so I needed to pass "matric" which gave access to the University and for this I needed to brush up my Latin and some scientific subjects.

Going back to school was not easy and it was made all the more difficult by the fact that physics, chemistry and mathematics were taught in a very dull way.

I never sat for the exam because in the middle of my studies I married Victor von Klarvill.

During the next two years I had a very easy time, but I did not find fulfilment in a life that was mainly social (even though it left me plenty of time for carving). I was therefore very happy when my husband decided to move to Kenya and suggested that I should go out and see what the country was like before we made the final break with Vienna.

In the event, the journey led to the break-up of our marriage. On the voyage out I met Peter Bally who was on his way to Kenya to look for a post as a botanist.

After my marriage to Peter I was confronted by a new world, the interest of which dwarfed everything I had known till then. Moreover it presented me with a challenge to take an active part in this world which was much bigger than I had ever dreamed of.

I had always been happiest in the country and particularly at Die Seifenmühle, but everything in Europe was dominated by man whereas in East Africa nature put man in his place and daily I saw around me wonders that surpassed all man-made inventions.

For the first five years, from 1938, I devoted myself to botany, although I did paint a few birds. I started by collecting and painting the indigenous flora as a hobby but after a while it became almost an obsession. Never before had I done any serious painting, but perhaps I had inherited a gift for it from my mother's family; certainly I found it easy to portray the exquisite colours and shapes of plants. Later a good many paintings were used as illustrations to books on East African flora.

When we were exploring the coral reefs of the Indian Ocean in about 1944 I became so fascinated by the fantastic colours and designs of the coral fish that I sketched them right on the reef before their bright colours faded. These paintings were bought by the Mombasa Municipality.

During the next ten years or so my interests gradually widened and I painted whatever took my fancy: insects, reptiles, shells. My ambition was to be as faithful as possible to my model but I did not want the result to resemble a colour photograph. For I believe that a work of art, even a minor one, should not only be a representation of the subject but also convey the artist's outlook.

In 1945 I became absorbed by the traditional customs and ornaments of the tribes of Kenya and I taught myself to paint portraits so that I could record the fast disappearing people and their activities.

Eventually the Government commissioned me to make a complete coverage of the most important tribes. In carrying this out I spent over six years living among these Africans, mostly in very isolated places. These pictures are now hanging in the State House as well as in the National Museum of Nairobi, and a few have been reproduced in my book *The Peoples of Kenya*.

I have tried my hand at landscape but I found this was not for me. Having trained myself to observe every detail of the subjects I painted I discovered that it was impossible for me to eliminate detail, yet this was essential if I were to convey the grandeur of Kenya's varied scenery.

I tried several times to return to wood-carving but most of my life was spent on safari in conditions which made such work too difficult. I only completed two tribal figures.

I had married George Adamson in 1944 and this has given me the chance of getting closer to wild animals than most people have had the privilege of doing. As George was in the Game Department we often found ourselves responsible for young animals that had been orphaned, and I was able to draw a few of these. But they turned out to be the most difficult subjects that I had so far attempted as they never kept still. Since an animal's life span is usually much shorter than man's, and since many of the orphans were brought to us because they were ill or injured, a number of them unfortunately did not survive for long. Nevertheless I have thought it worth recording exactly what happened in case the information can help others who find themselves in similar situations.

When Elsa and the other lions entered our lives ten years later, I concentrated on sketching them and this gave me a new and important way of understanding more about their characters.

I completed my animal sketches by drawing my cheetah Pippa and her cubs. I say completed because, sadly, I can no longer paint properly; owing to a motor accident, I have virtually lost the use of my right hand. The only painting I have recently attempted is of some Colobus monkeys that spent a lot of time in my garden at Elsamere.

This misfortune has however done nothing to prevent me from growing more and more interested in the behaviour of lions, or from working to save cheetah and other species that are in grave danger of being exterminated by man.

Now I am concentrating on helping the conservation of wild life and if thereby I can assist in resolving the existing conflict between the animal and the human world, then I shall feel that I am nearer than ever before to fulfilling my original quest – to find out more about the bewildering world around me – and that I shall be doing this in a constructive way.

1 THE FLORA OF KENYA

WHEN I ARRIVED FOR THE second time in Kenya in 1938, to marry Peter Bally, I joined an expedition to the Chyulu Hills organised by the staff of the Coryndon Museum (now the National Museum) in Nairobi. Peter was the Botanist in a scientific team which was to study the geology, fauna and flora of this hitherto unexplored volcanic range. I could not have asked for a more exceptional honeymoon, let alone for a more fascinating introduction to Kenya, than the following three months we spent in this twenty-eight-mile-long range of crater hills.

Craterostigma Lanceolatum scan

The area was of special interest because of its comparatively recent origin; as a result no humus had yet formed on the lapelli-covered slopes of the hills at its north-western end, and barely enough had yet collected in the craters to start a meagre vegetation. This condition gradually improved towards the south-eastern end of the range; here the hills were overgrown with such dense virgin forest that we had to cut our way step by step through a maze of creepers and lianas, and seldom got a glimpse of the sky in between the solid canopy of tree-foliage. Despite this profusion, there was no water on this range except for a tiny spring trickling from a rock at its north-western end. As we shifted our camp along the top of the range, we added more water to our scanty supply by sweeping sheets in the early mornings over the dew-wet grass, and wringing the soaked linen into a container. On rare occasions, we were able to supplement our store by finding a wild banana (*Musa*) tree in a stage of decay. In this state the normally sticky and bitter juice of the stalks is transformed into a crystal clear fresh liquid. From the stem and leaves of one wild banana plant we were able to press up to forty-eight gallons of water. This sometimes helped to solve our laundry problems.

Under such conditions, we had to be very economical with water, and I realised that the sooner I learned to respect it as one of the chief sources of survival, the quicker I would be able to adapt myself to my future life in Africa, which I was going to spend mostly on safari.

For the time being, I was entranced not only by living under canvas and so close to nature, but even more by having the opportunity to observe how plant life was beginning to grow on the almost barren north-western hills and gradually develop into lavish abundance towards the south-eastern end.

It was there that Peter opened my eyes to Kenya's botanical wealth which alone possesses five times as many species as the whole of Europe.

Being only too painfully aware of my unqualified status among the team of specialists we were with, I tried to make myself a little useful by starting to sketch flowers. On comparing my first effort with Peter's meticulous botanical painting, I was so disheartened that I tore the sketch up, but he patched the fragments together with scotchtape and encouraged me to carry on. I kept this memento of my first flower painting and I still have it. I shall always be grateful to Peter for having helped me develop a talent which up till then had remained dormant, for by doing so I was able to appreciate the unbelievably beautiful and varying flora of East Africa much more acutely than if I had just collected plants. (Later I did this as well for various international Botanical Institutes.)

After we returned to Nairobi, I was invited by Dr A. Jex-Blake and his wife Lady Muriel to show them my flower sketches. Both were pioneers of Kenya's horticultural development. Lady Muriel was well-known for her gardens, both in Kenya and in England; therefore I was very shy of letting them see my first sketches. But after they had seen them both the Jex-Blakes suggested that I should illustrate the second edition of *Gardening in East Africa*. I was truly alarmed. The book, which Dr Jex-Blake was editing, was to be a sequel to the first volume, with a similar title, and though it was a guide for amateurs on how to grow flowers and fruits in an African garden, there was a major emphasis on the indigenous flora and this was what they wanted me to illustrate.

It would involve the selecting of plants from various altitudes, from tropical coasts and arid plains, from luxuriant highlands and alpine mountain regions, always choosing the most representative species of each habitat.

Instead of feeling elated at being entrusted with this task, I felt more than reluctant to undertake it. I had hardly begun painting flowers, and what would happen if I used up the time allowed for these illustrations to be made and the result proved disappointing? In spite of my doubts, the Jex-Blakes persisted, we became very great friends, and during the next five years I illustrated seven books depicting the flowers, trees and

shrubs of East Africa. In the process I built up a collection of some seven hundred botanical paintings for the Coryndon Museum. Later, there was an exhibition of my flower paintings in London, for which I was awarded the Grenfell Gold Medal by the Royal Horticultural Society.

I was mainly interested in the beautiful flowering trees, of which East Africa has an unrivalled variety. Even in the arid regions, I found trees with such exquisite blossoms that one would never have thought it possible for them to survive in a semi-desert area. Many of them grow either foliage or flowers – but not both, since there is not enough moisture to nourish the two simultaneously. Others bloom only every few years for the same reason, and many will open their buds only after sunset, live during the cool hours of the night, and then wither away at the first glimmer of dawn. Some of these trees I was only able to find after dark and this by following their aromatic scent, by which they also attract insects for pollination. As many species blossom only for a few days, I had to be at the right spot at the right time to collect the flowers. I spent many nights trying to do justice to their frail beauty by the flicker of a safari lamp that attracted myriads of insects, which swarmed around me unless I worked under a mosquito net.

Later, when I climbed Kilimanjaro (just under 20,000 ft) and walked around the peaks of Mount Kenya (about 17,000 ft), I became so entranced with the alpine flora of these two extinct volcanoes that I returned and camped on Mount Kenya at an altitude of 12,000 ft to paint its unique flowers.

One can accurately call the flora of the East African mountains unique, for here, on the Equator, the plant life of the Northern and Southern hemispheres meet and one finds flowers that are common in the European Alps growing next to others found in South Africa. The most remarkable are the fantastic Giant Lobelias, which occur only here and (as a different species) in the South American Andes.

To get the best results, I chose the rainy season, which, at this altitude, meant snow. During the icy nights, the snow acted like an insulating blanket and protected the plants from the cold until noon when it thawed away and a strong sun warmed up the frozen ground. One can hardly imagine anything more beautiful than the early mornings, when the moorlands were covered by glistening powder snow, out of which the giant groundsel, the scarlet gladioli, the deep bluish-green delphiniums and the golden senecio stood up stiffly in striking contrast. At this high altitude, the colours of all plants were far more intense than at a lower level; the animals also were much larger here. I had with me a cook and gun-bearer, to protect me on my searches for plants from the elephant and buffalo which each morning came out of the dripping forest to feed on the moorlands.

Giant Lobelia

Ant-galls on Acacia

Protea Nadiense Oliv. 21

The most striking plants were the giant groundsels and giant lobelias. The latter, looking like candles in a candlestick, stood up to nine feet tall, in between the tussocks. They attracted species of sunbirds which, as they flitted around them sucking the nectar from the myriads of small flowers covering the surface of the hollow columns, looked like jewels. The birds took no harm from the caustic latex which nourishes the lobelia. On the other hand I soon discovered what disastrous effects its noxious vapours alone can have, when we carried a lobelia to camp because I could not paint it where it grew. By the time the gunbearer and I arrived with our load our lips and eye-lids were swollen, and we felt very sick. Even when next day I painted the plant from a distance of five yards, I felt nauseated. Later, I learned that people are also affected if they use the dried stumps of the lobelias as firewood when camping on the moorlands, and if they lick a drop of latex from their fingers they immediately vomit.

The tree-sized giant groundsel were too large to carry to my camp, so I had to paint them on the spot. After I had settled, with all my paraphernalia, in front of one near a small mountain stream, I concentrated so much on sketching the clusters of golden flowers which looked like huge praying hands reaching for the sky, that I was surprised by three elephants who suddenly appeared within a few yards of me, for a drink at the stream. Dropping everything, I ran!

Buffaloes also sometimes become a nuisance. Each night one bull in particular passed persistently through the narrow space between my mountain tent and the men's tent, brushing his body against the canvas. During these frightening moments, my little Cairn terrier, Pippin, who always slept at the foot of my bed, almost stopped breathing so as not to give himself away.

Despite such interludes, I found so many plants of immense interest that I extended my camping from one month to five, and only then packed up because the change in the weather began to make me rather jumpy. Each night I listened to the rhythmic waves of the monsoon as it came roaring down the moorlands from the ice-clad peaks, its volume increasing until it crashed with terrifying force against the forest. As the tents shook and creaked under the impact of the blasts sleep became impossible, and I expected to be hurled off the ground at any moment. So, with a heavy heart, I left this wonderfully eery mountain world—by this time I had begun to feel that I shared in its aloof detachment from the world below.

Having become captivated by the flowers of Mount Kenya, I naturally wanted to paint the flora of the other mountains too. So I joined a Swedish Expedition which was to study the wild life of all the mountains of East Africa.

We started with the Ruwenzori Range (16,794ft) on the Uganda-Congo border. These are

Senecio Brassica 23

24 *Merremia Ampelophylla* Hallier F

Gloriosa Virescens

the only non-volcanic mountains in East Africa, and have the reputation of being rained on for 360 days in the year. This may well be true, for within the forest belt such thick moss covers the branches of the trees that they look like teddy bears standing in deep layers of moss. Here walking is extremely trying and even on the moorlands we had to jump from tussock to tussock, otherwise we splashed into the morass below. In fact, we were very lucky with the weather during our seven weeks camping near the glaciers of the six mountains. The Ruwenzori range has often been identified with the fabled Mountains of the Moon to which the ancients attributed the source of the Nile. Ptolemy in his famous geography tells us that at a certain latitude "there rose the mountain of the moon whose snows feed the lakes, the sources of the Nile".

After leaving the Ruwenzori in a blizzard of snow, we spent four weeks in the crater of Elgon on the Kenya-Uganda border at an altitude of 14,167 ft. This extinct volcano has the most interesting flora of all the East African mountains. We pitched camp on a small island of very spongy ground, bordered on one side by an ice-cold brook, which lower down is known as the Turkwell River, and on the other by a boiling-hot spring. Between the two extremes we had no lack of water for every purpose, although during the nights we had to share it with the buffaloes.

When I finally parted from my Swedish friends, I had become so interested in the alpine flora that I continued to make use of every opportunity to paint and collect more specimens, so that to people who were not as lucky as I was in visiting this indescribably weird world, I could show, at least in pictures, the beautiful plant life of these mountains.

In 1944 I married George Adamson, and from then on my life was a continuous safari in the North Frontier District of Kenya where George was Senior Game Warden. This vast area of 120,000 square miles had hardly been explored botanically, so it was a challenge to collect and paint as many plants as I could. Of course, I had to adapt my hours of painting to George's job, and was often unable to finish a painting before we had to move to another place. Therefore I had to make the most of my time.

I well remember one morning when I was very keen to get a flower which I had found the previous evening close to camp, and which I hoped to paint while the tents were being struck. When I went to pick it, I found a herd of elephants at the very place where my flower grew, so I asked George to come and cover me with the rifle, hoping that the giants would allow me to collect my plant. But George was not to be hurried from his breakfast and after I had succeeded in making him move, he was not in a very favourable mood towards my artistic ambitions and followed me slowly.

When we arrived near the flower, the elephants were still holding the ground, and indeed a cow with a calf was within a few yards of the plant. I did not feel enthusiastic

Rubus volkansii Engl.

28 *Oncoba Routledgei*

Hibiscus Canabinus

30

about botanizing under the trunk of an elephant, but now George took his revenge on me for interfering with his breakfast, and he teased me in no uncertain terms about my failing courage. Just to show him that I was not going to be defeated, I walked boldly towards the flower, and picked it – but I was very quick in retreating to the safety of George's covering rifle! On the way back, he was remarkably polite – suspiciously so, I thought – and when we were safely in camp, he admitted sheepishly that he had forgotten to load the rifle! It was lucky that the elephants were so docile, and I was well rewarded for my involuntary courage when later I learned from the Royal Botanical Gardens at Kew that the plant was of exceptional interest.

I had another experience a few days later, which had less happy consequences. We had been riding ahead of the donkeys that were carrying the loads, and had chosen a shady tree for our lunch rest near a little rocky hill, which looked promising for succulents. To make good use of the time before the donkeys caught up with us and brought our food, I walked up the hill under the fierce midday sun. When I returned I developed a splitting headache. After lunch, we had to continue on our safari so as to reach the next waterhole before dark. Willy-nilly, I mounted my mule. The slow rhythm of its paces, together with the heat, made me still more sleepy, and I must have dozed off. I came back to consciousness by being thrown off the mule and landing with a very hard bang on a rock, while two rhinos crashed through the bush a few feet from me. The mule bolted. I was unable to move and was lying, groaning, on the rock-slab when George arrived. He had been riding ahead of me and when my mule passed him, riderless, followed by the two rhinos, he feared the worst, and was relieved to find that I was not dead, although I felt pretty close to it. The impact on the rock must have jarred my spine and paralysed me temporarily, for I could only howl inarticulately, and was unable to move. Unfortunately, not a drop of water was left, and the next waterhole was a good four hours march away. So with a lot of coaxing, I was lifted on to George's mule and somehow managed to cover the distance to the waterhole.

While camp was pitched I collapsed, exhausted, on to my camp-bed. Now the pain became acute. George tried his best to be a good nurse and to divert my interest from my aching bones, but somehow when he told me that this place was notorious for lions, and that on the last occasion when he had camped here he was looking for a man-eater who had devoured a Samburu herds-boy, I did not appreciate his bedtime story. George went on to tell me that he had put out a kill and, while waiting in his tent for the lion, must have fallen asleep, for next morning he found the kill still untouched but there were lots of lion pugmarks inside the tent near his bed.

This story at least made me want to get away as soon as possible – and, as I had broken

32 *Vellozzii* Sp. of Tomentosa Pax. Bak.

East African Crowned Crane
(*Balearica regulorum gibbericeps*)

no bones, we moved on again next morning. After what had happened I took great care not to fall asleep again while riding.

Even if the circumstances in which I painted were often rather too exciting, I enjoyed not only the sketching but also collecting dried specimens. By doing this, I learned a lot about the use of many of the plants for various purposes, such as insecticides, pigmentation, dyes, or food and medicine. I learned too what an essential place plants hold in the lives of the Africans – and reflected on the revival all over the world of herbal medicines and the research on new plants for the treatment of hitherto incurable diseases.

My collection of dried specimens involved working for many hours, usually after dark, pressing the plants I had collected during the day and preserving them from rotting during our long safaris. I also questioned the local Africans as to their use. Many were highly suspicious of my enquiries. But however exhausting my efforts often were I felt rewarded not only by the widening of my interests but also for being able, in a small way, to provide a greater knowledge of these plants.

Barleria Longissima

Edithcolea Grandis

36 *Hypericum Lanceolatum*

Regal Sunbird *(Cinnyris regius)*

2 A FEW BIRDS

THE BIRD LIFE OF KENYA—WHICH INCLUDES THE LARGEST number of species in any African country—presented a challenge to me. I wanted to paint some of these superbly-coloured specimens.

When I first arrived in the country, Dr V. L. G. Van Someren, a leading ornithologist and painter, was the curator of the Coryndon Museum. He had a remarkable collection of bird paintings—outstanding because of the iridescent effect of certain plumages he had conveyed, which many bird-artists have never caught. His secret was that he always used the purest watercolours and brushes that were almost one-hair-thin. For instance, to achieve the bluish-green metallic sheen of a starling, he would paint hair-fine lines of pure blue alternating with lines of pure green, carefully avoiding overlapping the colours. Thus he achieved an opalescent effect of unique quality. He never made use of opaque pigments.

REGAL SUNBIRD

Since I did not know enough about bird anatomy to tackle a living model, I started with a stuffed Regal Sunbird *(Cinnyris regius)* from the Museum. Trying to add line after line of different colours next to each other on the feather sections, which I had outlined, I soon found out that I could only do this with the help of a magnifying-glass. After concentrating for several days on the minutest details, my eyes got badly strained, and finally my vision was too blurred for me to continue. Unfortunately I was never allowed to watch Dr Van Someren at work—so I remained puzzled as to how he could have done so many paintings without becoming blind.

CROWNED CRANE

Next I tried my luck on a larger bird. I chose the East African Crowned Crane *(Balearica regulorum)*. This stately bird appealed to my artistic sense, not only because of its ornamental crown, but also because of its subtle colouring and exquisite shape. Again I found it too difficult to experiment with a living bird, so I used a mounted crane from the Museum. This was most frustrating, as I felt that all I achieved in the end was a lifeless study.

VERREAUX'S EAGLE OWL

Unfortunately, I never had enough time to study birds so as to be able to portray them well, and I therefore never made another effort to paint them, with the exception of sketching a Verreaux's Eagle Owl *(Bubo lacteus)*.

This happened quite by accident. One afternoon, when walking through the bush,

I found a large bird struggling frantically on the ground, flapping its wings and dragging one behind as if it were broken. It seemed to be in a state of utter exhaustion. When I approached very cautiously it attempted to get up, but collapsed. Now I noticed that one eye was badly injured, and that the surrounding fluff had been torn away. Judging by its smoky-grey down, it seemed to be a young bird, but even so it had talons the size of leopard's claws and an equally formidable beak.

As I did not feel brave enough to tackle this forbiddingly-armed bird alone, I asked the game scout, who always accompanied me on my walks, to return home and ask George to come with the Land-Rover and bring a strong cloth with which to rescue the bird. In the meantime, I sat within two feet of it; after a short while it opened its good eye and looked at me. I kept absolutely still and it gained confidence, gradually rising to a half-sitting position.

Soon after this, I heard the vibration of the approaching car, which seemed to horrify the owl as it fell to the ground feigning death. Then recovering a little it flopped about, staggering lamely, but kept its one eye fixed on me. When George arrived, he walked very slowly up to us, holding a hessian bag in his arm. As soon as he was within reach, the owl fell flat and remained so with closed eyes. It was a pitiful sight. Nevertheless, we took great care not to get scratched by the talons while we placed it in the bag.

Though I held the vulnerable load carefully on my lap during our drive home, the owl looked as dead as could be on our arrival. It seemed there was nothing left for us to do but bury it. George however suggested giving it a last chance to recover, so we placed it inside the wired enclosure which Elsa had occupied at night, and put a freshly-shot pigeon next to the owl. Then we left it alone.

After an hour we came back, only to find the bird, though lying in a different position, looking as dead as before. There was no sign of the pigeon. Smiling, George went to shoot another pigeon, which he left near the owl. The next time we approached the enclosure very quietly, and saw it tucking heartily in to the pigeon, but the moment it noticed us it dropped "dead" instantly.

From then onwards, the owl repeated the death-shamming act whenever a human being was in sight, even when we caught it clinging to the wire. Trying to find out how long it would remain in this uncomfortable position, I once sketched it whilst it was hanging almost upside down, but its determination was greater than mine and it kept up the faked death performance until I got sorry for it and went away.

It was interesting to see that many of the small birds which nested near the enclosure came fearlessly through the wire and drank from the waterbowl, in spite of the fact that the owl was sitting only a few feet away watching the visitors with one very open eye.

After two months the injured eye had gone completely blind, but it did not seem to hinder the owl's lively reactions and, as all the other ailments with which the cunning bird had deceived us so convincingly had been faked, we thought the time had come to release it. One morning we therefore opened the door to freedom, and, after a few hesitating flops, the owl took to its wings and disappeared from our sight.

40 *Lutzanus Sp.*

Theuthis Sp.

Thalassoma Sp.

41

Chaetodon setifer

3 CORAL FISH

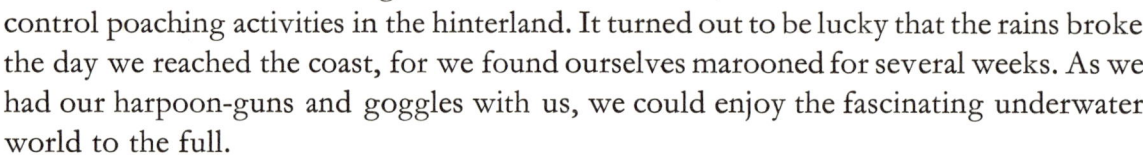

KIUNGA, A SMALL fishing village on the Indian Ocean at the Kenya-Somalia border, is a paradise for coral fish. We discovered this remote archipelago by accident during a safari we made so that George could control poaching activities in the hinterland. It turned out to be lucky that the rains broke the day we reached the coast, for we found ourselves marooned for several weeks. As we had our harpoon-guns and goggles with us, we could enjoy the fascinating underwater world to the full.

We soon made friends with Dilimua Bin Lali, an old fisherman of the local Bajun tribe, who took us daily in his flimsy dugout canoe to the reef. He told us which fish were good for eating, which were poisonous (he said they usually had some red marking as a danger signal), described the ones that were useful as bait, and others that were just *marodadi* (beautiful)–these of course interested me most, as I wanted to paint them.

Unfortunately, the coral fish lose their luminous pigments almost instantly when taken out of the sea, and I learned later that this is due to special cells *(kerai)* in the scales which dry up very quickly. I wanted to record their brilliant colouring and therefore took my watercolour paints out to the reef, and whenever we harpooned one of those fairy-tale fish, I placed it in a large *carai* (metal bowl) and sketched for two or three minutes before the pigments faded. Later, I hung the fish up by threads in a life-like position, and counted the exact numbers of fins and dorsal spines. I took great care to record the varying shapes of the scales of individual fish and filled in the details, using the coloured sketch I had made on the reef. In this way I made my fish portraits as authentic as I could. Often, when the tide came in late and we didn't return till after sunset, I had to sit up late into the night drawing our catch, since the fish would not keep their natural shape till morning. After I had finished painting them I salted the fish and thus preserved them until I could take my collection to the Coryndon Museum at Nairobi. There, plaster casts were made of the fish which were then painted according to my sketches–thus as near a replica as possible of the living fish was reproduced.

Apart from helping to build up the fish exhibits at the Museum, we had a marvellous time exploring the fantastic underwater world. There were so many coral gardens between the shore and reef that I don't think we managed to visit all of them during the many holidays we were later to spend at Kiunga. There was an infinite variety of fish, some with spots and stripes, others with squares and blotches; there were electric-greens and iridescent hues and so many eccentric shapes that we forgot time, hunger, cold, heat and sometimes danger, while gliding through this marine paradise. Sometimes we stalked a particularly interesting fish for hours, diving after it again and again until we were shivering from the cold of the deep water and had to give up.

On a few occasions, when we were too inquisitive, we got stung, or had to retreat quickly when we found ourselves suddenly face to face with a vicious-looking fish staring at us motionless, as if ready to attack. The black-and-white patterned Moray eels, especially, were often truculent and coiled up tight like a spring round the harpoon when we provoked them.

But most of these exotic fish played hide-and-seek with us in the fairyland scenery of valleys, canyons and mysterious tunnels formed by bizarre-shaped coral lumps inlaid with luminous rosettes or veined with fluorescent streaks of varying colours. All were overgrown with colourful seaweed which waved to the rhythm of the sea. Between the clefts and crevices peeped capricious coral fish, some shaped like golden boxes, others resembling butterflies. Often we could not see the body of the fish beneath its waving fringes or its elongated fins which floated behind it like a tail.

Swarming amongst the undulating tentacles of sea-anemones were tiny fish, which we could only catch by placing a mosquito net over the whole shoal and breaking off the coral block. Others, which were too small to shoot with the harpoon, we caught in basket-traps of narrow wickerwork baited with squid.

However helpful Dilimua was in finding the queerest fish for me to paint, he was merciless when he spotted the long antennae of a crayfish protruding from beneath a coral lump. Despite their formidable spiky shell, they were very easy to catch and kept completely motionless while we aimed the harpoon between their feelers. When they were landed Dilimua converted the old bailing-tin into a cooking-pot, in which he expertly roasted our catch. Crunching the juicy crayfish, still flavoured with the salty sea-water, we warmed up in the fierce sunshine until Dilimua discovered a new victim for our harpoons and we had to plunge into the sea, which could be icy at a lower depth.

During these holidays I had unknowingly acquired a unique collection, since nobody, up till then, had sketched the coral fish right on the Kenya reef. When I showed my paintings to marine authorities later, they found them so interesting that they wanted to preserve them. Fortunately, Sir Ali Bin Salem, a very charming and hospitable man, who was the head of the Kenya Coast community of Arabs, had endowed a fund some time before, and my pictures were bought with this money for the Municipality of Mombasa.

44 *Ostracion tuberculatus*

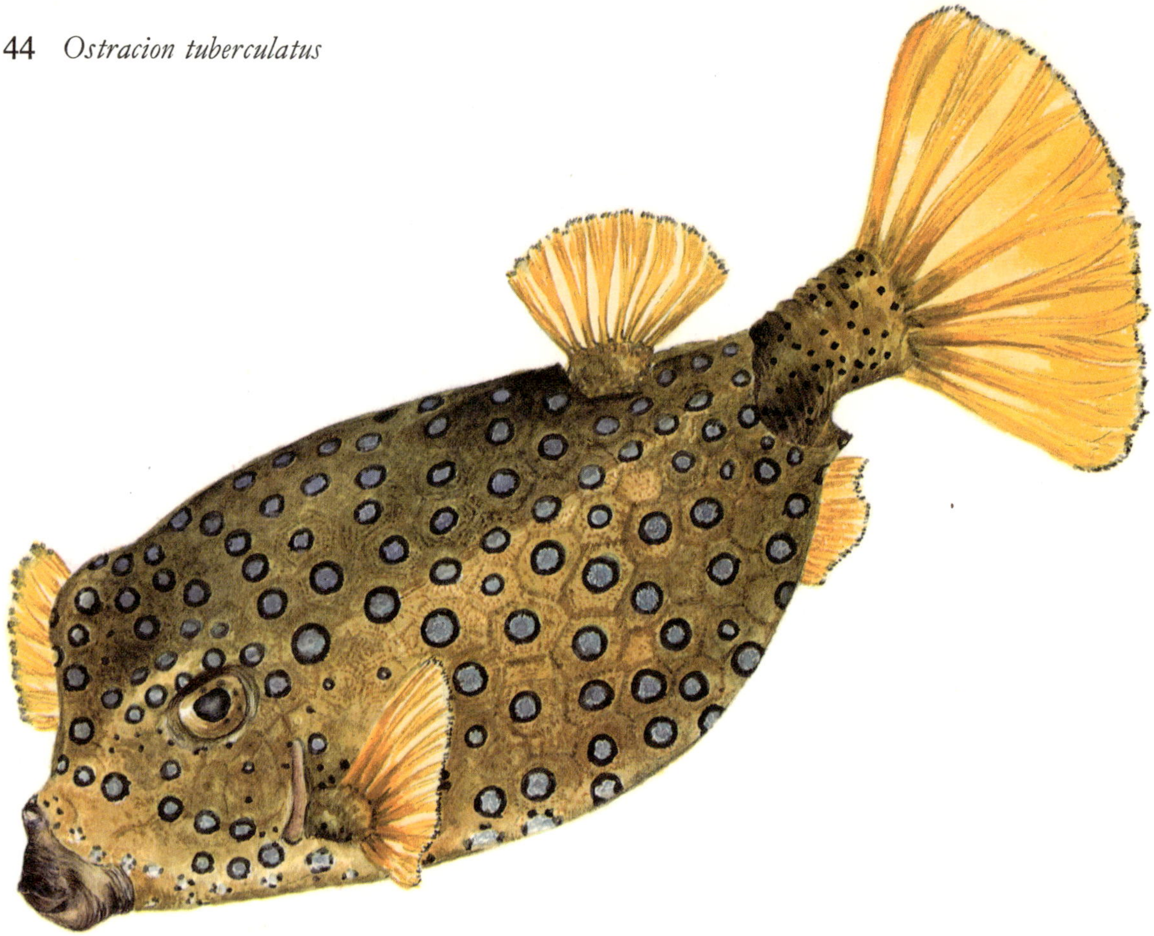

Balstapus aculeatus (Linn)

Holocanthus maculosos

4 WONDERS OF NATURE

HAVING PAINTED THE FLOWERS, BIRDS AND FISH OF KENYA I became interested in every other aspect of wild life and was intrigued by a variety of subjects which I can only describe as the wonders of nature and which included insects, shells and lizards.

A SIMULATED FLOWER

Some species in nature go to extraordinary lengths to protect themselves. In one case, which particularly fascinates me, what appears at first glance to be a flower spray is, on closer examination, a group of tree bugs *(Ityraea Gregoryi distant)* with folded wings, arranged along the stalk of a plant so that they simulate a flower. Sometimes bugs with pale yellowish-green colouring are found at the top of the stalk, and gradually darken in pigment the lower they settle, thus giving the impression of a withering flower.

I have never found this marvel of camouflage myself, but as I wanted to paint it and add it to my collection of Wonders of Nature, I borrowed a mounted specimen from the Museum.

The plant on which the bugs are here seen *(Achyrantheus aspera)* – the common burry weed – distributes its seeds by means of tiny, sharp hooks, which cling to anything that brushes against them (they often fastened themselves to my skin, and gave a painful prick if brushed off in the wrong direction). Thus they are transported by all kinds of involuntary hosts and, consequently, are very common.

This is what Robert Ardrey has to say about the flattid bug:

There is a creature native to Kenya called the flattid bug* . . . But to speak more precisely, what Dr Leakey introduced me to was a coral-coloured flower of a raceme sort, made up of many small blossoms like the aloe or hyacinth. Each blossom was of oblong shape, perhaps a centimetre long, which on close inspection turned out to be the wing of an insect. The colony clinging to a dead twig comprised the whole of a flower so real in its seeming that one could only expect from it the scent of spring.

My real moment of astonishment, however, was yet to come . . . The coral flower that the flattid bug imitates does not exist in nature . . . The flattid-bug society had created *the form.*

While I was suffering mental indigestion from the extraordinary statement, the eminent Kenyan . . . now contributed further material to my flattid-bug bewilderment. He told me that at his Corydon Museum they had bred generations of the little creatures. And from each batch of eggs

---

* *African Genesis* by Robert Ardrey, Collins Publishers, London, and Atheneum Publishers, New York, 1961

that the female lays there will always be at least one producing a creature with green wings, not coral, and several with wings of in-between shades.

I looked closely. At the tip of the insect flower was a single green bud. Behind it were half a dozen partially matured blossoms showing only strains of coral. Behind these on the twig crouched the full strength of flattid-bug society, all with wings of purest coral to complete the colony's creation and deceive the eyes of the hungriest of birds.

... Leakey shook the stick. The startled colony rose from its twig and filled the air with fluttering flattid bugs. They seemed no different in flight from any other swarm of moths that one encounters in the African bush. Then they returned to their twig. They alighted in no particular order and for an instant the twig was alive with the little creatures climbing over each other's shoulders in what seemed to be random movement. But the movement was not random. Shortly the twig was still and one beheld again the flower. The green leader had resumed his bud-like position with his vari-coloured companions just behind. The full-blown rank-and-file had resumed their accustomed places. A lovely coral flower that does not exist in nature had been created before my eyes.

STICK INSECT

This Giant Thorny Stick Insect *Palophus Regi* (Grandidier) I also borrowed from the Museum; the specimens I found myself were never as spectacular as the one I painted here. As long as the insect does not move, it is almost impossible to detect it amongst thorny bush, not only because of its colouring, but also because of the thorn-resembling growth all over its body. Its body is nine inches long, its overall length is fourteen to fifteen inches. Its wing-spread is about six inches.

COWRIE SHELL

There are forty-three species of cowrie along the Kenya coast, of which the tiger cowrie *Cypraea Tigris* is one of the most common and also one of the largest. Despite this, during many holidays spent exploring the coast, I only once found a shell with a living inhabitant inside it. This may be because the soft, jelly-like mantle of the living cowrie blends so perfectly with the sandy bottom of the sea that I may have passed one without spotting it. The tiger cowrie never lives on the reef but on the green grass-like weed that, in some places, covers the sand. When moving, the cowrie uses its mantle and the flexible feelers that protrude from it to propel itself. There are many theories about the use of the mantle but, so far, none is proven.

48 Tiger Cowrie Shell (*Cypraea Tigris*)

Pterois volitans

Flatid bugs (*Ityraea Gregoryi Distant*) clinging to the common burry weed (*Achyrantheus aspera*)

CHAMELEONS

I have a soft spot for chameleons because, for me, they represent the very epitome of "the exception to the rule" as far as lizards are concerned. To catch an insect they can project their club-like, sticky tongue for almost the length of their body and do this with lightning speed. They can move each eye independently; and have two fingers opposite the other three, with which to grip. They also have a prehensile tail and besides all this some species can lay eggs while others produce living young.

While the egg-laying sub-species bury up to forty eggs in the ground, and leave the incubation to warm climate, other sub-species give birth to what resemble inch-long parcels neatly wrapped in cellophane. We have watched a female about five inches long depositing, around a circle and with a foot-distance between each, thirty-six such "parcels" from which, a few minutes after being dropped, the young broke through and crawled about within their foot-long territory. Another female behaved similarly while producing twenty-eight young.

We all know of the chameleon's proverbial colour-change, but I only learned by accident one of the circumstances in which this happens abruptly. An American Zoologist, Sarah Atsatt, had just completed a year-long study on the chameleon, and was spending a few days at the Coryndon Museum, where we met. Demonstrating to us why chameleons change colour, she placed some in glass tanks, each tank covered with a different coloured tissue paper, but however bright the colours were, they had no effect on the chameleon. Then she banged on the glass and tried all sorts of noises, but with no results. Next she dipped the tanks alternately into hot and cold water, but these extreme temperatures did not stimulate any colour-change. Finally, she placed a living snake outside the tank, and instantly the pigments of the chameleon turned to their darkest hues.

While painting the *Chamaeleon Jacksoni (Vaurescecae)* (Kikuyu or Jackson's Three-Horned Chameleon), that I knew was a male because of his horns (which the female lacks), I learned of another peculiarity of these strange lizards. So that my model should not be able to move away during my sketching, I placed him on a stick which I hung in the right position by fine threads; along these the chameleon could not escape. While thus involuntarily posing for hours, he had to follow Nature's calls. In order to produce his faeces, he protruded a sac from his anus containing the digestive organs, which dangled for several minutes outside his body until the job was done. During this time the chameleon was utterly defenceless and, until he retracted the sac, vulnerable to his enemies.

I would like to mention here that the generally accepted slowness of a chameleon's movements is a fallacy: I lost countless of them, which I kept as pets, before realizing that they can move astonishingly quickly if they want to disappear.

In the case of this model, he had no chance to bolt – and became my travelling companion for several hundred miles, since I had hardly begun sketching him when George was called to Marsabit, 180 miles away, to deal with game offences. As the chameleon was a splendid specimen and I wanted to complete his portrait, I put him into a box with grass, provided him with enough live insects for the trip, and then continued my picture whilst camping at Marsabit. But before I had finished, we had to move on another two hundred miles, so the poor chameleon was once more on safari. Luckily, our destination was close to its original habitat, where we finally released it.

AGAMA LIZARD

This harmless, friendly lizard *Agama A. Lionotus* (Red-headed Agama Lizard) can also, like the chameleon, change its pigment to blend into the surroundings until it becomes almost invisible. This seems hardly possible with the strikingly-coloured male (the smaller female has a more protective brownish-mottled skin with dark red spots), but I can testify to it as I have sometimes only spotted the lizard by the movement of her eyes.

At the camp near my cheetah, Pippa, three pairs of lizards had almost become tame, and would bask on a fallen tree-trunk in the sun while I walked close to them, though at the slightest real danger, their heads bobbed up and down in rapid succession before they vanished into the hollow trunk. I often watched the pairs during their courting. The male bobbed his orange head vigorously while rushing after the female, who, curving her body into a tight arch, invited him to mate. After repeated bobbings and curvings, he finally mounted her; during this time both partners displayed their brightest colouring.

52 Giant Jackson's three-horned chameleon (*Chamaeleon Jacksoni Vanerescecae*)

Giant Thorny Stick Insect
(*Palophus Regi* Grandidier)

53

5 A RECORD OF TRIBAL TRADITIONS

IN 1945, IMMEDIATELY AFTER THE WAR, I HAD TO GO TO LONDON for several months on account of my health. To keep myself occupied during lonely evenings, I planned to carve a set of chessmen, of which the pawns would portray the various tribes of Kenya, the knight would be represented by a giraffe, and the castle by a phallic tower which we had found on ancient buildings along the coast—in short, I intended to make an ethnographic set.

To carve the figures accurately, I needed more detailed representations of the people I wanted to portray than photographs could show, and so, before leaving Kenya, I started sketching a few tribesmen. As I had no experience of portrait painting, I tried a different approach with each person. In one case I started with the nose and then grouped the rest of the features around it. In another I first outlined the head and then filled in the rest. Notwithstanding my trials and errors, I managed to get enough on to the paper to suit my purpose. Finally I bought chips of ivory from the Game Department and sawed them into the different sizes required for the varying types of chessmen.

When I began carving in London, I discovered that ivory is extremely hard and I then understood why Africans, before making their ornaments, soak the ivory for weeks in cow manure, so that the ammonia softens it. I copied their method in a more hygienic and concentrated way, only to find that the ivory cracked and lost its characteristic fatty texture. Then I tried a dentist's electric drill, but the result was disappointing, and I only completed the head of a giraffe.

56 Red-headed Agama Lizard
(*Agama A. Lionotus*)

Argonauta Hyans

Nudaurelia Cytheraea

Hoping to improve my portrait painting, I went for three months to the Slade School of Art. Although I recognised the privilege of being admitted for such a short time, I was not happy because I was obliged to concentrate on drawing the capitals of Greek columns, which it did not seem would help me to become more skilled in portrait painting.

Meanwhile, I had been advised to offer my sketches of the tribesmen, made for the chessmen, to the *Geographic Magazine*. To my surprise, they published them in colour in January 1948. I was still more surprised when the originals were later purchased by Sir John Ramsden, who gave them to the Coryndon Museum in Nairobi.

After my return to Kenya I wanted to paint more Africans, so I asked a friend to introduce me to several District Commissioners, in charge of tribes whose customs had been little interfered with by foreign influence. I hoped they might help me in getting sitters. I could not speak the vernacular of my models and, even with the few who understood Kiswahili, our conversation was limited to the number of their wives and children, livestock, and food. Many had never before sat so close to a European woman, let alone for such a long time, and they were tense and even frightened. To put them at ease I offered, apart from a financial reward, sweets and tobacco, and with gestures expressed my admiration for their ornaments and hair-styles. This worked well with the women, though I had to be more careful with the men who sometimes took my flatteries seriously; but on the whole I had very little trouble. To keep the expression of my sitters alive, I had to rely more on the silent language of the eyes than on any verbal communication, though this often diverted my concentration from my work.

To add to my difficulties my "studio" was usually under a tree and the light moved with the sun and shadows constantly changed. Also, occasional gusts of wind whipped dust over the painting, so I restricted my medium to water colours as they dried more quickly and were easy to transport.

In the beginning I was attracted by the Africans' picturesque apparel but I soon came to realise that their ornaments were not just decorations but emblems of their status, age and occupation. It therefore puzzled me why a few unrelated tribes wore similar ornaments. I had been especially struck when I found a pygmy woman at the base of the Ruwenzori on the Congo-Uganda border wearing a necklace almost identical with that of a Bantu man of the Wakamba tribe living near Nairobi, and also with that of a Gabra man of Hamitic stock living near the Kenya-Ethiopia frontier. Not only were these three races unrelated and of different cultures, but they now lived hundreds of miles apart. I wondered if the fact that they were wearing similar necklaces might provide a clue to their migration: and if, in the course of it, they might have met and exchanged certain rites and ornaments?

JELOJA GALAN
LEGATSHO, LAISAMIS
RENDILLE

SHEIKH AHMED BIN DAHMAN
HADRAMOUT – MALINDI
ARAB

I know that the most common test of racial relationship is based on linguistic similarities, but a comparison of "material culture" often complements the proof, and so it occurred to me that if I could provide such proof by painting all the tribes of Kenya at their various occupations and rites, I might help to find out more about some still obscure migration routes. Such a record might also be of immediate value since many of the tribes of Kenya were already replacing their traditional customs and ornaments with Western garb and soon their apparel would be forgotten unless it could be preserved in pictures. But though this idea was a challenge to me, I had neither the funds to carry out such a task, which it would take many years to accomplish, nor the authority to win the co-operation of the Africans to allow me to paint their initiation and other secret rites – unless Government would support me. However, since I was not a professional painter, nor an anthropologist, I did not feel justified in asking for such support.

Nevertheless, I was so interested in sketching whatever models I could find that I went on painting. At this time, I was staying at Kapsabet with my friend, John Thorpe, who administered the Nandi. He was building a welfare hall in honour of those Nandi who had distinguished themselves during the Burma campaign or had been killed during the war. When John asked me if I would make a life-size oil painting of a Nandi in Burma kit to be hung up in the new hall, I felt very embarrassed. I had never painted in oil before, let alone undertaken such a large picture. It was to show a soldier down to the waist, in his war accoutrements. In spite of my doubts John persuaded me to try; we agreed that he would be under no obligation to buy the painting if it turned out badly, but that he would pay me a fee of £15 if I were successful.

Sharing the Thorpes' guesthouse was a Government official whom I had not yet met as he was on safari. He returned while I was converting the verandah into a studio and trying to arrange my sitter with all his war kit in the best position. As an ex-Army officer, he was able to help me to place the rifle correctly, see that the medals were in the correct order, and that the wide-brimmed hat was cocked at the proper angle.

After I had drawn the outline of my sitter and my new friend had unpacked his kit, we talked under a loquat tree; we were both eating the juicy fruit and competed to see which of us could spit the slippery pips the furthest! He became interested in my painting, as well as in my idea of making a record of the Kenya tribes in their traditional costumes.

When he introduced himself as Pat Williams, Commissioner of Social Welfare at the Ministry for African Affairs, and offered to take my small collection of paintings to Nairobi to show to the Governor and members of the Legislative Council, in the hope that they might commission me to go ahead, I agreed. So, while Pat and I carried on our loquat-pip game, my dream became a reality, and my life for the next six years was decided. Little did I know that from now on I would be travelling into the most remote parts of Kenya to paint warriors, dancers, witch-doctors, sorcerers, prophets, rain-makers, blacksmiths, elders, brides and widows, and even such secret rites as circumcision, and that eventually I would paint seven hundred portraits in which the traditional way of life of Kenya's fifty-four most important tribes was recorded.

In my book *The Peoples of Kenya* I have described in detail my experiences during this period. Today about two hundred of my best portraits are hanging in The State House, and most of the rest are exhibited at the National Museum in Nairobi.

Apart from making an historical record, I was mainly interested in catching the character of my sitters so that I should not only produce an ethnographical dossier but also a collection of living people. I tried to achieve this by holding the attention of my sitters with my eyes and thus automatically, but unconsciously, I believe I was able to convey the spirit of my models. In our "silent eye language", we both communicated in a far more direct way than words could have achieved. Sometimes this was a great strain on me and had an almost hypnotic effect on both of us.

I remember especially the time when I was painting *Laibon* Koney Ole Sendeo (a *laibon* is a spiritual leader of the Nandi and Masai tribes, whose authority is based on mystical powers and who acts as an intermediary between man and the other world). Koney Ole Sendeo was a direct descendent of one of the most famous *laibon* of the Masai, after whom one of Mount Kenya's peaks is named. The ritual make-up of a *laibon* includes white circles painted round the eyes. These gave Koney Ole Sendeo a strange, almost inhuman expression. To make matters worse, he attempted to mesmerize me, thus rendering it even more difficult for me to concentrate on painting his eyes. When he realised my difficulty he suddenly remarked, "I am not suprised that you cannot get my eyes on to your paper, because they are not normal eyes." Determined to disprove this bold statement, not with words but by my painting, I carried on until I was satisfied with his portrait.

Luckily not many of my models presented such a challenge. But some whom I would have liked to have painted refused to sit for me because they believed that I would then possess their spirit and gain power over them. In a few instances, I convinced the person that this would only be the case if we looked into each other's eyes, and that if he posed in profile I could not harm his spirit.

Trying to improve my painting in order to convey the rich pigments of the African complexion, I developed a technique of my own. I usually started with the highlights and went into darker tones by putting one layer of paint on top of another layer until I

achieved the effect I wanted. I almost modelled in paint to get the strong contrast of the bone structures.

Since many of my sitters had to come from far distances, I worked out a schedule and allowed two days for each person, provided they turned up in the early morning: as this seldom happened, I was often obliged to paint the many strings of beads and other ornaments at night after previously outlining their position. Sometimes the colouring of the beads was not to my taste, but since they were strictly traditional and indicative of the person's rank, I had no alternative but to copy exactly what I saw. My wish to vary the pictures in design and format was rendered impossible partly because the size of the paper had to be standardised and, partly because I was obliged to show off the ornaments to their best advantage. But in the end the ornaments, which symbolised the status of my sitters, proved so fantastic that they outweighed all these problems.

It was interesting to see the reaction of the Africans to their portraits. The majority were baffled by what I was doing and after viewing their picture from every angle, admitted that they could not understand a human head represented in two dimensions. Many of them, however, could carve three-dimensional figures. I was all the more proud when a few appreciated their portrait and offered me their livestock or other prized possessions if I would give it to them.

I was astounded to learn from a witch-doctor what seemed to me to be the basic concept of impressionism. He carried a ritual wooden staff with the figures of a man and a woman at its top. While he had carved the woman in detail, the man lacked the characteristic male anatomy though he made up for this by his extremely arrogant expression. When I asked why he had not completed the figure of the man, the witch-doctor imitated the stupid, vacant gaze of the woman, and then the man's superior look. Obviously for him these were far more indicative of the difference between the sexes than their anatomy. He had also made a few drawings symbolising the spirits of various illnesses which he was frequently asked to cure. They were outlined in human figures but the thorax, abdomen, and extremities were broken up into geometric designs, similar to ones I have found on carved wooden effigies along the coast. I was perplexed by these superb expressions of anatomical representation in an abstract form by these apparently primitive people, especially when compared to some of the modern abstract art in civilized countries.

Of all the various forms of art in which man has expressed himself since the earliest ages, I value the Egyptian culture most, for it is my belief it achieved an unsurpassed combination of abstract and naturalistic interpretation of what the artist wanted to convey.

6 SOME ANIMALS

ROCK HYRAX

We were camping within a few miles of the Equator, at an altitude of about 8,000 ft. While we were having breakfast, a local tribesman approached our tent, clutching a tiny animal in his hand.

When I inspected the little creature, which was so young that its umbilical cord was not yet healed, I didn't guess that I was facing someone who was to become my most intimate friend for the next six-and-a-half years, but I soon saw that it was a female hyrax, about one or two days old.

I knew nothing about hyrax then except that they are considered, because of their teeth and feet, to be the nearest relation to rhino and elephant and that they are mentioned in the Bible under the name coney. I also knew that there are two species. One is arboreal and nocturnal, its habits are solitary; this species is known as the tree hyrax *(Dendro hyrax)*, which is notorious for its ear-splitting screams which shake the forest during the night: the other species *(Hyrax procavia brucei)* is diurnal and lives in colonies among rocks, it is commonly called a rock hyrax or rock rabbit, or, in South Africa, a dassie *(Hyrax procavia capenis)*.

Since the new arrival was too young for me to identify its species, I assumed she might be a tree hyrax, probably because these were the only hyrax I had seen (in captivity). Remembering their vocal reputation, but wishing to be polite, I named the little animal Pati Pati, after the Primadonna, Adelina Patti. However, to my relief, I soon discovered that she had no inclination to entertain us after sunset, but fell fast asleep as soon as daylight faded and never made a sound until morning. But by the time I knew that she was a rock hyrax, I did not like to change a name that seemed to suit her so well and Pati Pati she remained.

I tried to feed her with diluted, unsweetened Ideal milk using improvised teats made of chamois leather, but for three days Pati obstinately refused food. Finally we succeeded in getting milk down her by using a hypodermic syringe with a metal mouthpiece and thereafter for eight months Pati greedily drank enormous quantities of Ideal milk. Of course in the wild she would have been weaned much earlier, but she had become so fond of "her bottle" that she refused solid food long after her teeth were ready to deal with it.

Because Pati was only a day or two old when she joined us, she at once adapted herself to our life. During the day she spent most of the time on my shoulder and at night she curled round my neck. I soon got used to wearing this little, living muff.

All hyrax have a gland on their spine which shows up as a light-coloured patch about two inches long. When the animal is excited, either by pleasure or alarm, this exudes a sticky substance which has no smell; at the same time their hair stands up.

Pati was the cleanest creature I have ever known so I was puzzled that she was constantly scratching herself with the claw that hyraxes have on the inner toe of each hind foot. Later, I learned that this is called the "toilet claw" and is used for tidying up. In many ways, hyrax are an interesting species, and still puzzle zoologists. Although they are the size of guinea-pigs, fossils have been found in Egypt of varieties as big as a horse. Its molar teeth serve to classify the hyrax as belonging to the great group of ungulates or hoofed animals, and thus – with one exception – their toes are armed with broad nails that suggest small hooves. The exception is a remarkable one: the innermost or second toe is armed not with a hoof, but with a long curved claw (the toilet claw I have referred to) thus recalling one of the features of lemurs. This resemblance may be of significance for the lemur seems to be connected, at least in fossil form with the ungulates. There are other features too which suggest that the hyrax descend from a very primitive type of mammal, which may not have been far removed from the parent form of the Anthropoid.

Alternatively, if one considers this group of sub-ungulates, the hyrax may have links with the elephant, the sea-cow, dugong and manatee. The distribution of hyrax is confined to Africa and the eastern Mediterranean.

Altogether, there are fourteen different species of hyrax, all of which have feet similar to those of rhino or tapir. They have a further peculiarity, which is that the naked sole is furrowed; this enables the hyrax to cling to the vertical sides of rocks and trees. Their teeth are like those of rhino, in particular they have a single pair of rootless, upper

incisors which grow from persistent pulp, and keep on growing; their two pairs of lower front teeth are both rooted; and their seven molars are separated from the incisors by a large space.

Hyrax have an abnormally large number of vertebrae – twenty-eight to thirty – compared with man's twenty-six. They lack a developed collar bone, have only a rudimentary tail, no gall bladder, and a cleft upper lip. Their eyes are also different from those of other mammals in that they can move their nictitating membrane. The stomach recalls those of horses and rhinos – the placenta is zonary, as in elephants and carnivores. Most species of hyrax have three pairs of teats, one of which is placed near the fore limbs, and the others situated to the rear. But in three species of tree hyrax, there is but a single pair of teats.

However bewildering a creature the hyrax may be, I valued Pati more because of her remarkable character than her zoological peculiarities. She would do many things just to please me, but nothing was ever gained by trying to force her. Pati had no defences other than her very fast reactions, unbelievable courage, and her formidable upper incisors. She was house-trained by nature. In the wild she would have used a rock, always the same one until it became too dirty, from which she would have let her droppings fall. (Rhino have similar habits, but do without the rock.)

Living with us, she adapted her habits and replaced the cliff by the throne of the W.C. Here she perched herself on the seat, evacuating her droppings tidily into the water below. However convenient this habit of hers was, it confronted us with a problem when we were on safari as then we had no W.C. Pati simply would not function without water below her, so we had to improvise a substitute, using an old filter to which George added a wooden seat.

Hyrax are vegetarians, and Pati was particularly fond of any plant containing latex (though to many animals latex is poisonous) fig-leaves and euphorbia – also liked by elephants – became her favourite food. Although she fed most of the day on plants which grew wild, she never missed a chance to hop on to the table opposite my plate and share my meals as well. The starchy foods, such as bread, potatoes, rice and porridge, she found especially delicious, but bananas, eggs, paw-paws and mangoes also provoked a tug-of-war between us.

Pati's greatest weakness was "drink" – she was a toper. Of course we tried to control this bad habit, but Pati was extremely cunning, and whenever she discovered a bottle of spirits, she would topple it over, give the cork a pull with her teeth, suck the spirits out and go to town.

She shared everything with us, and of course joined us on all our safaris. When we were on foot safaris, mounted on mules, Pati had to ride on my shoulder. At the beginning it was a bit tricky getting the two animals used to each other, but the mule soon accepted Pati's efforts to hop along its back to find a cooler spot than my shoulder offered, and tolerated her gymnastics. On the other hand, Pati clung most cleverly whenever, on a narrow path, an unexpected jostling from the mule bumped us against a tree or a rock.

Several times Pati was obliged to travel on a dhow and had to endure sprays of water splashing over her when the sea was rough. She was well known along the coast and often visited the fishermen in their dark huts which were close to where we camped. They gave her tit-bits, dates and cashew nuts.

But of all the safaris we made together, Pati enjoyed our trips to Mount Kenya best of all. There on the moorlands, at an altitude of about 12,000 ft, she was truly in her element. For us, it was most interesting to see a deep-rufus coloured hyrax, of a much larger size than Pati, amongst the rocks at about 15,000 ft altitude. Obviously a rock hyrax, for there were no trees except a few giant groundsels which were of no use to a hyrax.

Pati had plenty of opportunities of meeting wild hyraxes of her own kind, but strangely enough she never showed the slightest interest in them and always remained indifferent to her kin.

Pati was about a year old when a very young hyrax was brought to me. I hoped she would adopt it, but her only reaction was jealousy and I had to free the little one; otherwise she might have killed it.

When I was commissioned by the Government to paint the Kenya tribesmen in their traditional ornaments, Pati proved an ideal companion, and for months on end she was often my only friend. She was also a great help with my models; while I painted she would sleep in my lap or rest on my shoulder and observing her kept my sitter's expression far more alive than it would otherwise have been.

Pati was five years old when George returned from hunting for a man-eating lion with three tiny, newly-born cubs, who were unfortunately orphaned by this hunt. Pati at once adopted them, and with touching affection never let them out of her sight. Whenever possible, she nestled herself close to her charges, which they seemed to like. Indeed they accepted her as an equal, but soon the three little lionesses began teasing Pati by pushing their soft noses under her tummy to turn her over; which she tolerated admirably. Then the day came when the cubs out-grew their "nanny" and Pati, realising their superior strength, had a hard time combating their playful attacks, for she had no defence other than facing the little rascals with such an air of authority that they left her alone.

Pati had always been Number One in our household until the arrival of the cubs. After this a great change took place, for most of our visitors were so captivated by the attractive little lions that Pati often went unnoticed. I was touched by her reaction. She never seemed jealous of her rivals, remained very dignified, and when I cuddled her, she ground her teeth – which was her way of showing that she was happy.

Unfortunately we could not indefinitely keep three fast-growing lionesses living in freedom with us, and so, very reluctantly, we sent two of our foster children to the Zoo in Blydorp, Rotterdam, when they were five months old, and kept the smallest of them – Elsa.

Pati and Elsa had such different ways of life that they never interfered with each other and both animals got an equal share of our affection, and had their special privileges.

It was remarkable how, as members of our family, the lioness and the hyrax respected each other during the many safaris we all made together.

When Pati was six and a half we noticed that she was getting old and less active than she had been. Then, on a safari to the coast, she began to be badly affected by the heat. I protected her carefully from the sun, but she suddenly had a heat stroke and became very ill and, within a few hours, died in my arms. It took me many years to recover from the loss of my friend.

LEOPARDS

A leopard, which had been giving trouble killing livestock, was trapped and arrived one morning at Isiolo. The plan was that George should release him where he could do no harm. As far as we could judge, the terrified beast, which was turning round and round inside a small wooden crate, was a beautiful, full-grown leopard.

While George went off to find a suitable location where he could liberate him, I used the opportunity to sketch him. I had never been so close to a leopard before, and was fascinated by his emerald-green eyes. I did not then know that they only turn this colour when the animal is in a frenzy. Slowly I walked to within a few yards of the crate and sat still, waiting for the animal to calm down, but he went on throwing himself against the wire door, snarling and growling at my slightest movement.

Suddenly I had the idea of growling back. I imitated his deep sound as best as I could. This seemed to puzzle the leopard who instantly grew silent. He stared at me with murderous hate and backed inch by inch into the farthest corner of the crate, never taking his eyes off me. There he crouched and watched, listening to my growls. After a while, I stopped. At once the leopard shot forward to the door and in a fury, tried to break through. Again I growled and again he became quiet and retreated into the back of the crate where he kept still as long as I continued my growls. Gradually I changed these into a soft purr which seemed to have a soothing effect on him. But the moment I stopped, he again charged snarling at the door. If I wished to sketch him, I had to go on purring. In the event our duet continued until George's return.

By then it was late in the afternoon, and this was the best time to let the creature go. So, at the base of a cliff near a little spring, we cautiously opened the door to freedom and watched the leopard disappear towards the setting sun. It was a very happy moment for us all.

How much the poor animal must have hated every moment of that day I only realised many years later when I adopted little Taga, a tiny leopard cub about two weeks old. She was so young that most of her skin was still pink, and her rosettes so densely set together that her fluffy fur looked almost black. I could not have wished for a more affectionate, gentle and highly intelligent companion than Taga, about whom I have written a chapter in *The Spotted Sphinx*. During the time we were together she never once showed the faintest shimmer of green in her deep amber-coloured eyes, which looked so soft when she was happy. Like Pati, she sometimes slept curled around my neck where she evidently felt warm and secure. In answer to my stroking, she uttered a tiny sound.

Unfortunately Taga developed tick-fever when she was six weeks old, and died of it.

TWO BABY ELEPHANTS

One day we received a message that a baby elephant, abandoned by its mother, had fallen into a shallow well. It had been rescued and now urgently needed a home. Would we look after it? Of course we would! It was arranged that a Land-Rover should collect the little elephant from some sixty miles away and bring it to us next morning.

Hurriedly we made preparations. Luckily we still had the big, open enclosure in which we had kept Elsa at night, where he would be safe from the carniverous animals that often roamed around the place. The enclosure was filled with sand and large slabs of stone, to which we added knee-deep dry grass to make it soft and warm. I phoned several people throughout Kenya who had had experience in rearing elephants, and asked their advice. Following their suggestions, we bought tins of glucose and acquired remedies to deal with pneumonia and stomach upsets, both of which were likely consequences of the little elephant having spent a whole night in the bottom of a well. We

knew that elephants, in spite of their robust appearance, are the most difficult animals to rear, particularly when very young, so we wrote to the Uganda Game Department, which has made a special study of them, for advice.

Full of excitement, I awaited the arrival of our new charge. It was the first time we had taken on a baby elephant, and it was a great responsibility. I decided to call him Pampo. When he arrived and was lifted out from the car, the volume of his scream might have put an adult elephant to shame.

I tried to comfort Pampo. He was very strong, although his stomach and ears were still baby pink; we guessed that he was not more than two to three weeks old. His skin was badly chafed above the eyes and pelvic bones, where he had grazed himself against the rough walls of the hut in which he had spent the previous night. He accepted the bottle of warm water and glucose which I gave him to quench his thirst. This was going to be his diet until his diarrhoea had cleared up. As he seemed very hot and was breathing irregularly, I dosed him with Sulphathiazole against pneumonia.

Then I sat down in the enclosure, and very soon poor little Pampo rested next to me, utterly exhausted. No wonder he was tired, after all he had gone through: first the shock of falling into the well and losing his mother; then, bewildered, he must have spent a dreadful time in the cold water, drinking probably more of the dirty liquid than was good for him; and this had been followed by the ordeal of being rescued and locked in a hut for a lonely night. Finally there had been the long journey standing in a Land-Rover, bumping over bad tracks, until he reached us.

Pampo had every reason to bear a grudge against the world, and I was therefore all the more touched to find him keeping trustily as close to me as he could and placing his heavy head on to my lap. I stroked his rough and deeply wrinkled skin, searching for soft places which I found under his lower jaw and on his shoulders. The latter were protected by his enormous ears which folded close to the body and kept the skin underneath slightly damp. The inner side of the ears was exquisitely patterned like a giant leaf with a network of veins. Towards the edges they were as thin as parchment and slightly undulated. They have a thermostatic effect. I took his tiny trunk in my hands; it was so soft and flexible, it seemed incredible that it should have such a concentration of muscle underneath the skin. At its extremity were two protuberances which would serve later as very sensitive "fingers"*, but at present Pampo had little control over them. His lips were slightly hairy and protected a tiny, very pointed tongue. Where his trunk and forehead joined, the skin was deeply folded, and when relaxed the overlapping segments

* The Indian elephant only has one.

were like a closed concertina. The massive forehead towered over the powerful eye-socket bones and already acted as a miniature bulldozer. How often we had watched adult elephants pressing their foreheads against the stems of doum palms until they toppled over and crashed to the ground, bringing the desirable nuts within reach.

The hide on Pampo's pillar-like legs was hard and thick, especially around the forelegs, and already well suited to withstand the toughest conditions. But the soles of his feet were still tender and vulnerable to the smallest thorn. Knowing that newly born calves walk great distances with the herd, this puzzled me. I was fascinated by the complicated pattern of deep lines all over his body, designed to allow free movements of the huge joints underneath. Even if one knew nothing of the bone structure of elephants, by following the purposeful wrinkles and folds leading up to what looked like seams, from which they branch off in different directions, one could guess at the animal's anatomy. If adult elephants are sometimes compared to relics of a bygone age, this seemed still more relevant to Pampo, whose deeply furrowed face and expressive eyes looked as though he already knew all about past and future lives.

There is a period of about twenty-two to twenty-four months gestation before an elephant is born. At birth he weighs around 230 to 250 lb. He will not be mature before he is twenty years old and may live up to eighty. The first teeth break through within the second month, pushing their way forward along the gums. When their surfaces get worn out at the front, they are gradually replaced from the back. There are no canines, and the incisors develop into the tusks. The teats are between the front legs and there are only two of them. I well remember watching a cow suckling her calf and lifting up one front leg to give it better access. Elephants differ in their foetal development from more recently evolved mammals. While we–being the youngest species–show our evolution in embryo, by the presence of gills and tail, it is a long time before we look like a *Homo sapiens*; being a very old species the elephant resembles an elephant right from the beginning. During a post-mortem of a female which we had had to shoot, we found an embryo of about three months; already at this early stage it was well proportioned and looked like a miniature elephant. On the other hand its toe-nails, its rectangular lower jaw, a tiny lump behind its head and its tusks are not at this stage characteristic of an elephant but of a mammoth, the ancestor of the modern elephant.

My thoughts about elephants were soon interrupted by urgent problems. Pampo needed constant company throughout the day and the night. We decided to engage an African to sit with him while I had a break. We also brought in a young sheep as night companion. We called her Silly and she was a great success. The sheep and the elephant accepted each other at once; Silly trotted everywhere behind her big friend and slept close to him at night, and apparently did not mind being a little crushed at times.

Finding an elephant-sitter was not so easy, as he had to meet with Pampo's approval, and Pampo was very selective. After several trials, we found the right companion and it was an idyllic scene to see all three–Pampo, Silly and the youth–sleeping close together during their siesta.

After three days nursing, Pampo seemed to have got over his pneumonia and his stomach upset. Soon he was drinking up to two gallons of cow's milk a day. Into this I mixed glucose, cod liver oil, boiled rice water, and a little salt. The little bull was rather particular about the temperature, and every three hours the bottles had to be warmed up to body temperature; apart from this we had no trouble in feeding him, though it was difficult to procure enough milk in the dry season.

In addition to reasons of cleanliness it is essential for elephants to have a daily bath, in order to keep their skins moist and healthy. As soon as Pampo showed how fit he was by enjoying long walks and feeding well, we gave him a mud bath twice a day. As water was scarce, every drop having to be transported three miles, we compromised by splashing a bucket over him, which he did not much like. After this we encouraged him to roll in the warm sand which he did with squeals of delight. He took some time to get his heavy body into the right position; with great care he would lower himself on to his knees before the rest of his body collapsed. But often the position was not quite to his liking. Then up he would struggle, straightening his front legs and, when he had found exactly the right spot to roll in, the strenuous procedure of lying down was repeated. He loved it when I rubbed sand well into his skin and this kept him clean and free from vermin. Afterwards he would doze off happily, lying on his side.

There is argument about the position in which elephant sleep. As a rule we have seen them standing under the shade of a tree, during the midday heat, often leaning up against the trunk, but also, on rare occasions, stretched sideways on the ground. Each time it has surprised me how quickly these huge animals can rise and bring their six-odd tons under control. Pampo confirmed the story that elephant will often use a termite hill as a pillow; he always tried to keep his head at a higher level than the rest of his body, and often put it on my lap or on a stone slab. When satisfactorily settled he would snore contentedly.

After a few days we enlarged his night enclosure to include the base of a big, rough-barked tree so as to give him an opportunity of rubbing himself against it. Up to now, he had often used us for this purpose, and whenever his rough hide scratched against our skin it felt as if we were being rasped with a cheese-grater. In order to keep the enclosure clean we engaged another African to cut fresh grass, for we knew that the slightest fouling of Pampo's sleeping quarters might have severe consequences to his health.

Within a fortnight, Pampo lost his pale pink colouring and developed a shiny blue-black fluff all over his body. He needed a lot of exercise and we walked daily from 4 p.m., when the sun was no longer hot, until dark, at a good pace – Silly always came too, bleating along some distance behind. Pampo loved his walks, and whenever I stopped to give him a rest, he just pushed me aside and went on.

During the first few days he never let me out of his touch, and persistently either stepped on my heels or fondled my hand with his trunk as we went along. But soon he learned to walk alone, often stopping to rub a thorn off the tender soles of his feet. On such occasions he was always willing to let me help. I wondered what happened to an elephant in the wild state? Can the mother extract a thorn with her trunk?

It was not long before Pampo discovered convenient gullies where he could roll happily in the warm sand, but he still found it difficult to recover his balance and bring his bulky body back into an upright position. He also had to learn to overcome ob-

stacles, such as stepping over fallen branches or going round them. However fast he could race up the few steps leading into our house, he was nervous of soft ground and most reluctant to descend the slightest sandy slope. Considering the astonishing agility of adult elephants in climbing the steepest hills and negotiating the most broken ground, his behaviour showed that this is not an innate skill but has to be acquired by learning.

Pampo was very intelligent and persistent. If he wanted something, he persevered until he got it. He was most determined about a ground-sheet with which we covered the windward side of the enclosure at night. He spent hours chewing the hard edge of the old canvas. We imagined it helped to ease his teething pains, although as yet there were no signs of any teeth. One day, someone told us that tent canvas is often impregnated with an arsenical preservative. Ours was bleached by age, nevertheless we removed it as a precaution, and replaced it with a heavy piece of cloth hung on sticks for him to chew at. Pampo, shaking his head disapprovingly, tried to get at the old canvas which was now out of his reach and even stood on his hind legs to do so.

He had a great variety of sounds to express his emotions. When happy, he would squeal or make a deep burring noise, and would curl his little trunk affectionately around my neck, or hold my hand with it. Sometimes I blew down his trunk, then he would retaliate by sending a blast of spray into my face. He seemed to like this game and when I laughed he would repeat his blasts without further provocation from me. When protesting against his shower-bath, he would make a touchingly gentle noise and would try to run away. We thought his aversion to water might have originated from his experience in the well and we hoped he would get over it. Pampo never screamed again as he did on his arrival.

When on the defensive, he would retreat several paces backwards to give himself momentum for a charge, but often he did this without provocation, or just to show that he had had enough milk and wanted no more.

He was very conservative in his habits as regards feeding-time, sleeping and exercise. After two weeks with us he tried picking up small objects between the two protruding ends of his trunk, and tried to push grass into his mouth. But whatever he did, he needed company and even during the nights, when he seemed fast asleep, he would wake at once if Silly moved away.

For three weeks, Pampo seemed to be a happy and rapidly growing elephant. Then I noticed that he was getting thinner in the face, and seemed less energetic. His consumption of milk dropped from two gallons a day to only six bottles. As he was continually rubbing his gums against anything he could find, we thought that teething pains might be responsible for his loss of condition and appetite. His affection was touching, and he seemed only to be at ease when I sat with him and let him rest his head on my lap.

One day he took a great interest in his bucket, and pushing his head right into it, sucked up the water, nothing could stop him. We took this as a sign that he would soon be weaned from bottle-feeding and drink from a bowl, which would greatly facilitate feeding him. Next day he wanted to repeat the performance, but as he had a slightly upset stomach we refused to let him have the bucket. Soon afterwards, we found him trying to satisfy his thirst from a muddy puddle below a waste-water pipe. Daily his digestion trouble became worse so we called in the vet. He advised us to stop all milk and feed only glucose and water mixed with sulphaguanadine to clear his stomach. From now on, poor Pampo became progressively weaker and soon refused even this food. Achromycin failed to stop a violent diarrhoea which produced egg-like lumps of undigested milk.

Early one morning, exactly four weeks after his arrival, I found little Pampo in a deep coma. I called the vet at once, and we tried to revive him with brandy. It seemed to ease him and his gasps became less frequent. Three hours later his condition showed a slight improvement—his breathing was calm, and his heart was again beating steadily—so the vet went home. I sat close to Pampo; he responded to my stroking by looking at me—he had not moved from the position I had found him in at dawn. We looked intently at each other. While I tried to hold him to life by my love I felt there was a current between us—he responded with his eyes. I rested my head close to his so as to hear any change in his breathing, and placed my hand above his heart. It was throbbing feebly but steadily. We remained like this till suddenly I realized there was no more breathing and no heartbeat. I had not noticed when both had ceased. Pampo was dead.

We helped the vet to carry out a post-mortem. It showed that the poor little elephant had acute enteritis in his intestines and stomach, aggravated by ulcers. He also had nephritis and severe pleurisy. It seemed incredible that with all these illnesses Pampo could have survived as long as he did, and been so active. All records from the Uganda Game Department, the Rome Zoo and the Rotterdam Zoo showed that in spite of the most modern scientific care, less than 10 per cent of young elephant calves survive in captivity. Analysis reveals that elephant milk contains 3–4 per cent protein, 4–6 per cent lactose, 6–7 per cent fat, and 0.4 per cent minerals. Moreover, it has two and a half times more Vitamin B, and four times more Vitamin C than cow's milk; it has never yet been successfully substituted.

Although the findings of the post-mortem explained poor Pampo's death, they did not console me for his loss. He had been such a lovable animal. About a month after Pampo died, we were asked to look after another baby elephant. It had fallen into the same well, and had also been abandoned by its mother. It was another little bull—but much younger than Pampo had been—still very pale in colour, and pathetically thin. We called him Podo.

As soon as I walked up to him he charged with all his might, stepping a few yards backwards and rushing at me and repeating his attacks. I was told that he had not eaten all day and had refused everything except water, which he had sucked greedily from a bath-tub.

We had learned through our experience with Pampo that cow's milk would be fatal, and we therefore followed the advice of experts by trying powder-milk mixed with glucose, Farex, white of eggs, and a little streptoguanadine, to control his bowels. But whenever I came into sight with the bottle containing this mixture, he shook his head violently and even several times pushed the bottle out of my hand. Then I tried unsweetened Ideal milk, diluted but without other ingredients; this he liked better. But in spite of all my efforts, he drank so little of the milk that I became desperate. On the other hand he drank quantities of water from the bucket, into which he often dipped his head as well as his front legs and at which he sucked and sucked many times a day. I tried to substitute diluted Ideal milk for the water — but he was not to be fooled and would not go near the bucket unless it was filled only with plain water. To aggravate our difficulty, injections failed to stop his increasing diarrhoea.

Then I thought of one last way of trying to get nourishment into our stubborn charge. He seemed to have a passion for sucking the brass knob-handle of the kitchen door, in fact he spent most of his days pressing his head against it. I had tried holding the milk-bottle in its place, but it was impossible to keep it in any position from which an elephant could suck. We therefore unhinged a similar door, placed the brass knob at a convenient height for Podo to suck from, and carved a hole underneath the knob through which, from the back of the door, I inserted the rubber teat of the milk bottle into his mouth whenever he sucked the knob. This trick seemed to work — at least I was able to make Podo swallow more nourishment, though not enough to sustain him. So we wandered about with the mobile door — an African holding it in position, while I hid behind it with the milk bottle. The elephant nearly pushed both of us over in his efforts to suck. It was a cumbersome way of nursing.

In spite of this device, he became thinner and weaker, and after a few days his legs were no longer able to support his weight and we had to lift him into every position. I spent all day and all night with him, placing my camp bed next to his straw-padded corner in the large enclosure. But although the vet and I tried our best to save Podo, he died after ten days.

From the post-mortem we found that the main trouble had again been very acute enteritis, but though there were no signs of pleurisy or pneumonia, as we had found with Pampo, Podo's lungs also adhered to his ribs. Since all the connecting tissues looked so healthy, we wondered if this adhesion was not natural in elephants — though I had never heard of it occurring in other mammals.

In Pampo's case I had believed this to be a result of his pleurisy and pneumonia, but as there was no trace of either in Podo, I was puzzled about this adhesion, and wrote to Osman Hill, a world authority on elephants, who replied: "The adhesion of lungs to chest wall is normal in the elephant and develops during the second half of foetal life; hence it is already established in the new-born animal. A similar condition is found in hippopotamuses and probably occurs in other large pachyderms."

AN IMPALA

We had been following vultures expecting them to guide us to a "kill", when they led us to an Impala doe lying defenceless on her side in the open sun. She was terrified when she saw us, and tried to get up but was too weak to rise. Meanwhile, more and more vultures circled above her, ready for their gruesome task.

Our first thought was to provide shade for the poor doe, so we cut branches and arched them over her. Then we approached her very slowly, intending to examine her wounds. She had none; her only symptoms were a very blown-up tummy and a swollen and bleeding vagina. We assumed that the trouble was probably due to a pregnancy or a miscarriage, but though I made as much of an internal examination as I could, I found nothing to account for her condition except that her muscles inside were very tight and the passage was septic. All I could think of doing to relieve her was to massage her tummy and so gradually reduce the gas inside.

The young doe seemed to know that we wanted to help her, for though she had kicked when we first came, she now lay still and even co-operated. I was touched by the trust she showed by allowing us to touch her after such a short time. After four hours of gentle massaging, her tummy got softer and smaller and her legs, which had been stiff, were relaxed. But the grim audience of vultures was still growing and since we could neither leave her to a horrible fate nor hope to cure her quickly, we tried to lift her into the Land-Rover. This made her kick so violently that we were obliged to give up the attempt. However, we were determined to move her, so next we tried to get her on to her feet by doubling her legs under her and lifting her body into the sitting position which antelopes use when resting. Then we gently raised her hind legs, and to our great relief she at once got to her feet and raced away. After struggling about twenty yards, she collapsed. We gave her time to recover and repeated our efforts. This time she reacted in the same way and ran about twice the distance before she fell. A third attempt resulted in her racing off and then taking up a normal pace. After a few hundred yards she disappeared into the bush. We were very happy for her sake, but kept looking in the direction she had taken to see whether the vultures were following her, and soon we saw them descending and felt sure that she had collapsed. We found her lying on the ground, very exhausted, so we decided that this time, kicking or no kicking, we would lift her into the car and take her back the six-odd miles to our camp.

After padding the floor of the Land-Rover with grass, I sat in the back with her, in order to control her if she struggled; but she put me to shame, for she made no attempt to kick and behaved as though she were quite accustomed to travelling by car.

In camp my truck was waiting, it had a wired-in back, which provided an ideal nursing home. We covered its floor with a thick layer of grass and lifted the poor doe into it. She soon relaxed and could not have been more gentle.

For the next four days I treated her with sulphathiazole mixed with unsweetened Ideal milk and cod liver oil. She drank it well and her digestion functioned normally. We propped her up inside the car in a sitting position, leaning against the side panel, so as to keep her in as normal a posture as possible, and every day we tried to put her on her feet. But she was too weak and after a few minutes she started to tremble, then her legs gave way. All the same, we went on making the attempt, so that she should not lose the use of her legs, and also to keep her spirits up. She seemed to recognise that George and I were trying to help her, for she showed no fear of us; on the contrary, she seemed to like being stroked and held her head in a position which showed that she approved of our handling her. But when any other person came to help change her grass bed, she barked at them, clearly differentiating between us.

I spent most of my time with her, caressing, feeding and sketching her. For three days I hoped she might recover. The bleeding was less, her appetite was good, her digestion normal, she seemed relaxed and satisfied, but she was still very weak. On the fourth day she refused food, and soon it appeared that she could no longer swallow. This condition lasted for the next three days. Whenever I tried to drop some milk into the pouch of her lips, she dribbled it out at once. As the days passed, her bones became more and more visible and the colouring of her reddish coat less vivid. Defenceless, she looked at me with eyes that were now very large. It made me miserable not to be able to help this lovable and exquisite creature. I comforted her as best I could, but by now I knew that I could not save her. I hardly dared to leave her for an hour.

Exactly one week after we had found her I was sitting beside her, sketching when she suddenly gave a moan and opened her velvet lips. A cramp-like shiver went through her until she became rigid, her neck arched backwards as far as it could reach and her body trembled in the last effort of her life which was now ebbing slowly away.

She died so gently that I could not distinguish when life left her, even though I was waiting for it with my hand over her heart. She lay there, as beautiful in death as she had been in life. She was such a delicate and helpless animal who had given us her trust and contrary to her normal reaction to humans, relied on receiving the best from us, while she herself, by her very nature, gave of her best.

Although I could not save her life, I felt at least grateful that I was allowed to help her die peacefully and unmolested.

ARABIAN ORYX

In 1962 one of the most difficult and costly rescue operations in the history of animal conservation was carried out to prevent the extermination of the Arabian oryx (*Oryx Leucoryx*). There were less than two hundred of these wild animals surviving in Arabia, and these were threatened by motorised hunting, against which the oryx had no chance. Often as many as three hundred vehicles at a time took part in a hunt. In 1960, after a report by Lee Talbot about the alarming situation, the Fauna Preservation Society decided to organise Operation Oryx in the Eastern Aden Protectorate to secure enough animals to form a nucleus for a breeding herd. Major Ian Grimwood, Chief Game Warden of Kenya, was put in charge of this expedition for which he sacrificed his overseas leave.

With only two men to help him, he camped for many weeks in the desert, endeavouring to get near the elusive oryx which, having been mown down by machine-guns, kept well away from the sight of a vehicle, let alone of men. Despite incredible difficulties and risks, Grimwood's team succeeded in capturing two males and one female which were brought to Kenya. They were kept at Isiolo which has a similar climate to that of the Eastern Aden Protectorate. For several months the oryx were confined in large pens at the veterinary quarantine station. This was within walking distance of our house. George helped supervise the feeding and care, and I naturally watched and sketched the oryx whenever I had an opportunity of doing so.

I was intrigued to see that the Arabian oryx were of a colour that blended with the desert, thus providing camouflage, whereas the two species living in Kenya – the *Oryx beisa annectens*, and the fringe-eared *Oryx callotis*, are of a dark reddish-grey colour. The Arabian oryx is the smallest of all oryx and the delicate, subtle hues of their coat make them seem more frail than their Kenyan cousins. In spite of their benign appearance, because of their rapier, sharp-pointed horns, all oryx are formidable opponents, even to lions. When seen from a position in which one horn hides the other the oryx might be credited with having inspired the legend of the unicorn. In one of the many stories woven round the origin of this imaginary creature in relation to the lion, it is presented as fierce enough to rival the king of beasts in competition for the crown. An old nursery rhyme describes their feud:

> *The lion and the unicorn*
> *Were fighting for the crown;*
> *The lion chased the unicorn*
> *All around the town.*

Then the story goes on to describe how the unicorn chased the lion across the skies. When finally the lion returned to earth he dodged behind a tree. The unicorn, by mistake, rammed its horn into the trunk and thus became an easy prey for the lion. This was an inevitable end to the tale; since then the lion is the king of beasts. Today, the unicorn is the emblem of Scotland, and together with the lion, supports the Royal Arms of Britain.

The three oryx at Isiolo adapted themselves surprisingly well to captivity, even though they were occasionally visited by lions (who however were outside the enclosure).

Several establishments in various parts of the world were approached to see if they could provide a secure breeding station for oryx; not only for those at Isiolo, but also for some owned by Sheikh Jabir Abdullah Al Sabah of Kuwait, and King Saud of Saudi Arabia.

In 1962, I was invited to Philadelphia to show the film George and I had made about Elsa, at the final ceremony of the 100th anniversary celebrations of the Academy of Natural Sciences. Being rather nervous of my first public appearance in the USA (at which I had also been asked to address several receptions in connection with the launching of the World Wildlife Fund in New York), the Elsa Wild Animal Appeal invited

Major Grimwood to accompany me so that he could give more detailed information on conservation projects in Kenya than I could provide. After we had completed our tour, Major Grimwood flew on to Arizona to negotiate with the Maytag Zoo at Phoenix for a possible home for the oryx where they could breed. When an agreement with this Zoo had been reached, the Shikar Club, USA, helped to finance the flight of the three oryx from Isiolo to Phoenix, where they were joined by a few others, given by private owners.

In 1964, during my second lecture tour in the USA, I visited the herd and was very happy to find my old friends from Isiolo in excellent condition, and even happier to watch the first born fawn gambolling on strong legs after its mother. Simultaneously a breeding herd had been established at Riyadh and another at Slamy. By 1972 the herd at the Phoenix Zoo numbered thirty and that at Slamy thirty-three. New herds at Los Angeles and Abu Dhabi had eight and two oryx respectively.

It is hoped that these breeding herds will eventually produce a large enough number of oryx to rehabilitate the species in Arabia; it is also hoped that National Parks will have been established there, where the oryx will be safe from machine-gunning.

A YOUNG BUFFALO

We were on a foot safari near the Ethiopian border and had climbed with our pack donkeys and mules up Mount Kulal. The isolated position of this extinct volcano, which rises from the surrounding desert at 1,100 ft to an altitude of 7,000 ft, attracts the clouds and often its upper region is wrapped in thick mist while intense heat simmers above its lower slopes. In consequence, Kulal has a rich vegetation of virgin forest with dense undergrowth at its summit, and has become a refuge for many animals, especially the greater kudu and buffalo. Unfortunately, this was well known to the tribesmen of the area, and George had come here to investigate poaching.

One day, his game scouts brought into camp a baby buffalo, whose mother had been killed. It was a tiny bull, at the utmost one month old. He staggered into camp on long, knobby legs, looking pathetically thin. On his head two small, rounded knobs were just visible; later they would grow into massive horns. Most of his deep grey skin, including his long, tapering, soft ears which ended in a pointed, hairy fluff, was covered with long hair. In Samburu language Long Ear is "Segeria Egitok" – and this is what we named him.

But what were we to do with this pathetic creature? We intended to stay only a few days on Kulal before traversing the desert country below. As Egitok was used to the climate of Kulal and cool forest shade, we could not take him along on this foot safari through hot country. George suggested that the kindest thing would be to shoot him. I looked at him; it was impossible. Was there no alternative? Suddenly I had an idea. Why not try and make him ride on a camel across the desert until he reached our lorry some 150 miles away; he could then be transported to a friend who had a farm at the foot of Mount Kenya where the climate was similar to that of Kulal.

Our friend was interested in finding out whether wild buffalo could be crossed with

92 Impala

Bushbuck

domestic cattle, but up to now he had failed in his experiments as he had only had female buffaloes. If our little bull could survive the trip, we might not only save his life, but also perhaps throw some further light on the crossing of these bovines. At least he would have the advantage of being brought up with cattle. As Egitok had nothing to lose, we decided to give him this chance.

Another fact in favour of my scheme was that George already had many poachers in our retinue waiting to be sent to prison. They were a heavy drain on our food resources. Egitok's journey would be an excellent opportunity to send them all together to Baragoi where our lorry was waiting. What we needed were camels as transport for Egitok.

We told the Samburu tribesmen to bring some milk every day for the little buffalo, and explained that we also wanted to hire a cow in milk and two camels and a herdsman for a trip to Baragoi. In the meantime, we hoped to feed up Egitok so that he would be strong enough for this adventure.

He took greedily to the bottles of cow's milk, and within a few days had enough strength to toss us all in the air. It was a favourite game of his. As soon as he saw somebody off guard, for instance bending over a box of some saddlery, he would approach as silently as he could from the rear, and, pushing his head quickly between the legs of his victim, give such a jerk that the unsuspecting person turned a somersault and landed on the ground. The onlookers found it very funny and everybody laughed hilariously whenever he repeated this performance. This seemed to encourage Egitok, for he lost no opportunity of playing his trick.

In spite of his rough horse-play, our staff loved him and competed in offering him the best sleeping place at night around the fire. I was touched when I saw our men lying

uncomfortably far from the warm flames so as to leave the best position for the little buffalo. One scout especially, who had a squint eye, loved him dearly and I often caught him unaware, stroking his friend and picking ticks off his thick skin.

As I always fed Egitok, he was naturally particularly attached to me and followed me everywhere. Whenever we moved camp he trotted good-naturedly behind me for long distances, sucking at my thumb which was the only lead I ever used.

On our walks, while exploring Kulal without Egitok, we often met buffaloes. As a rule we had to break our way through dense undergrowth across steep slopes, or along very narrow ridges, and when meeting these huge beasts on a precarious path the question arose as to which of us would turn away first? I was often scared by these colossal creatures, who stood facing us, their solid helmets and sharp, pointed horns raised in a suspicious attitude before, luckily, they decided to turn round.

After two weeks on Kulal, Egitok had become very tame and was easy to handle. He had also filled out and so we thought that we could risk sending him across the semi-desert country to Baragoi. The Samburu had not only brought a milk cow with her calf, but also two camels which they wanted in any case to send to Baragoi, so all worked out well for our buffalo. Even the problem of how to secure Egitok on to the camel was more easily solved than we had expected.

On their nomadic wanderings the Samburu use a kind of wicker basket as a packsaddle. It is made of four wooden frames interlaced with rawhide thongs and then hinged together, two at each side, forming panniers. Into these they pile their possessions, tying them on firmly with ropes of hide.

We padded a wicker frame with grass, and fastened Egitok so that he could lie down

as in a cradle. Then all was mounted on to the protesting and gurgling camel. Squint-Eye was put in charge, and to his encouraging "ttoe-ttoe", the strange caravan started to move. First the spare camel as leader, tied tail to head with the one which carried Egitok, next the cow, as the living milk bottle, with her calf, and finally, a string of prisoners and their escorts. We watched them going down the mountain until they were out of sight. What an odd procession it was.

Many of the poachers were on their way to gaol because of their offences in hunting buffalo; what an irony, I thought, that they should be led to their punishment by a little buffalo whose mother had been their victim.

After a few weeks, we rejoined the party in Baragoi, where everybody gave us a tremendous welcome. All were proud of Egitok, who had become the poachers' pet. He looked fit and was very lively, although he had been a little chafed by the thongs around his ankles, but these abrasions soon healed.

When finally we presented our friend with the little buffalo, he was more than pleased.

7 LIONS AND CHEETAHS

During the years in which the lioness Elsa and later the cheetah Pippa lived with us I had unique opportunities for drawing these splendid creatures.

Knowing them as well as I did after spending many hours every day with them while they played, fed, ate or slept, I was soon able to recognise their every change of mood.

This was due not only to the way in which they held their lips, dilated their nostrils or tilted their ears and heads, but even more to their general expression and the look in their eyes. When they were relaxed their eyes were soft and warm but at the slightest alarm these changed in a flash and became hard, murderous even; the pupils reduced to pin-head size. On the other hand, when they were roused to a state of great emotion, for instance when alarmed about their cubs, the pupils became so dilated that the fine amber-coloured iris was reduced to a circle and looked like a thin frame that held the deep black cavity in place.

I have often wondered how the change from the round pupils which is characteristic of wild felines to the slit pupils of domestic cats evolved.

I found Elsa much easier to sketch than Pippa. This was probably due to the difference in character between the two species. Lions, which are to me the epitome of dignity, are very self-assured; they are also lazy, whereas cheetah are very highly strung and always on the alert; to me they represent the quintessence of elegance.

Both Elsa and Pippa were very affectionate but they showed this in different ways. Elsa demonstrated her emotions openly, perhaps because she belonged to a gregarious species, while Pippa, belonging to a solitary species, was much more restrained and hated to show her affection if other people were present.

However something they had in common was the great care they took not to scratch me when we were playing together, and if by accident this happened and they drew blood they always seemed to be much more upset than I was.

When I held one of their paws in my hand it was so trustful that I could hardly imagine that this sensitive velvet paw could rapidly change into an effective killing weapon.

Like all lions Elsa used her front paws to hold her food in position but Pippa, in cheetah fashion, bent her forelegs close to her body while gnawing at her meat.

Another trait they shared was their marked dislike for being sketched. Perhaps they felt that when I did this I regarded them more like models than as friends?

I could understand their discomfort for having posed several times for a portrait, and painted so many myself I was well aware of the strain on both model and artist.

When I was being painted I usually knew, without having to look at the painter or the canvas, just what part of my anatomy was being portrayed and often felt as though I were being mentally dissected. Since the senses of animals are more acute than ours it is reasonable to suppose that in the same circumstances they are under even greater stress than a human being.

Anyway, Elsa and Pippa certainly hated being sketched and as a result usually turned their heads or hid them behind a paw or walked away when they saw what I was doing. For this reason most of my drawings show them asleep, eating or occupied in some other way. Only very rarely could I portray them looking at me.

Because I wanted to catch their expression and their spontaneous movements I never finished a sketch after the animal had changed its position.

I regarded these sketches chiefly as an aid to observing the animals and recording graphically, in greater detail than I could by looking at them, what they were like.

But I had planned to make finished pictures from the sketches later on. However, in 1970 I had a car accident in which my right hand was injured, and I can no longer paint. Nevertheless I feel that my drawings contributed to a closer knowledge and understanding of the lovable characters of Elsa, Pippa and their cubs.

8 A COLOBUS FAMILY

WHEN, IN 1970, I MOVED TO LAKE NAIVASHA I HAD NO IDEA that a pair of colobus monkeys *(Colobus Guereza Kikuyuensis)* were living there. It was a rather unlikely place for them as the altitude is only 6,200 ft and these arboreal monkeys normally prefer a higher habitat.

Perhaps they were on my land because it includes a small area of indigenous forest which provided them with food and with good shelter in which to hide.

It is possible that they were descendants of the many monkeys who inhabited the forest around the lake before it was reduced in size. To be able to vanish was essential because, owing to their striking pelts and the fact that they make good eating, colobus have been mercilessly hunted. Indeed so tragic was the state of these monkeys that today they are protected, but this unfortunately does not put a stop to the activities of poachers.

The colobus are by far the most beautiful of all monkeys and when they jump from tree to tree with their long black and white cape trailing behind them, caught in a shaft of sunlight, they look more like fairy animals than real creatures.

Once as George and I were watching the pair jumping among the acacia trees in front of the house, a branch broke and one of the monkeys landed on the ground with a terrific thump. It had fallen from a height of at least one hundred feet and I felt very worried thinking that it must have injured itself. I searched everywhere, but for the next ten days we never saw a sign of the pair.

When at last they reappeared I watched them and at once noticed a pale patch which I took to be a bruise at the end of the female's arm. Then I collected my field-glasses and had another look and to my joy discovered that the pale patch was the face of a pure white baby monkey sheltering under its mother's protecting cape.

So far as I could judge, it was about a foot long including a tiny tail. The father colobus stayed near to his family, watching out for danger, while the mother suckled the baby and when she moved she clutched it to her with one hand while she hung on to the branches with the other.

We named the baby Coli. From now on I was able to watch the family almost daily. The father was always on guard until Coli grew a little older and would crawl clumsily around as his mother rested her head on her mate's lap. Mother and infant were so inseparable that they appeared to be only one body. The monkeys fed in the early morning and after four o'clock in the afternoon; they slept during the midday hours and again from dusk to dawn.

The juicy leaves of a senecio climber seemed to be their favourite food but they also enjoyed munching the berries and the young shoots of some pepper trees which had, long ago, been brought to Lake Naivasha. For the rest they ate the leaves of the thick undergrowth.

When Coli was two months old his hands and feet turned black and his arms, face and head went grey but leaving a thin white line which divided the brow from the rest of the head and connected the white side and front-neck.

When his mother leapt from tree to tree Coli would cling tightly to her belly. They remained inseparable and showed great affection, gazing at each other for long spells while embracing each other. The mother groomed her baby constantly and while this went on he clasped her round the neck. So far as I observed Coli never touched his father.

As the days passed the family became quite relaxed in my presence. I could approach within about thirty yards of them but they always remained at least thirty feet above the ground. The father was much more nervous of me than was the mother, who evidently trusted me for she even let me film her with little Coli on her lap. Strangers alarmed them very much and often if visitors had been to the house they would not be seen again for several days.

By the time he was three months old Coli had developed a little tassel on the end of his tail, which was still white. Now, he was able to jump round his parents or, while holding on to his mother, to grab a small branch from which he would dangle helplessly until he managed to somersault backwards on to her. She, for her part, allowed him to hop off and on to her head with unbelievable patience.

It was soon after this that Coli's father began to play a useful part in his education: sitting two feet away from his mate his shoulders provided a splendid spring board from which Coli could leap on to his mother's shoulders. In this way the little monkey learned to jump small distances in safety. His parents co-operated further by widening the distances between them, so teaching Coli to judge how far he could safely jump. They also usually kept their backs at a comfortable angle from which Coli could easily slide down. His mother in particular would stay for long periods hunched up, but if Coli didn't do his exercises properly then she would spank him.

The drill went on in complete silence and indeed colobus monkeys are quiet creatures, but when they do give voice it is with a beautiful, deep, sonorous call like a roar and unlike the sound of any other monkey except the howler.

Coli's father, when he was about to call, usually started by making a clicking sound with his tongue, then holding his body low he would jerk his head at each roar and this would go on for five or ten minutes. The mother often joined in but with less volume and did not go on calling for so long as her mate. When Coli was frightened he just squeaked and, on hearing this sound, whichever parent was nearest to him rushed to his rescue.

By the time he was five months old Coli, though still very fluffy, had developed the characteristic colouring of an adult.*

By now he was independent in his movements except that he could not climb trees with a trunk too thick to allow him to get a grip. To help him one of his parents would sit in a fork above him and dangle its tail down to within his reach. Grabbing it, Coli would heave himself up and thereupon the parent would move up to a higher fork and repeat the performance.

There were times when I observed him defeated by the size of a trunk and, lacking a tail to help him up, go to another thinner tree and climb to its top from which he would jump to the thicker tree. On these occasions his father would sometimes climb up ahead of him and, when the two met, Coli would use his father's shoulders as a springboard from which he would slither down to his mother who was waiting below.

Coli was a very busy little monkey, always twiddling round while hanging from a small branch with one hand or hopping along to swing from another tree. He was also a tease and would sometimes snatch a leaf from out of one of his parents' mouths. For this he always got the spanking he deserved. Although I always thought of him as a male it was not until Coli was sixteen months old that I could be absolutely sure of his sex.

It didn't take him long to know when I wanted to take photographs of him and he was far from co-operative. As soon as he saw me coming with my camera he would sneak off, hide behind a tree from which he would peep at me only to withdraw again. As soon as I lost patience and turned away, he would come into view and perform the most startling acrobatic tricks and the expression on his face seemed to say "Now take my photo if you can". But if I returned he vanished at once. How I loved this little monkey and his parents.

When Coli was thirteen months old I had to go to the Meru Park for a stay of nearly three weeks. I was sad to leave the happy and inseparable colobus family even for this short time, but I wanted to try to see Pippa's cubs.

On my return I was told that for the last two weeks only Coli and one of his parents had been seen. I spent a whole day searching for the missing monkey; then I saw it high up in a tree close to my boundary. He seemed to be holding onto a branch but he did not move when I came close. Suddenly I realised that it was the father and that he was dead.

The bark of the acacia tree in which he was sitting was so smooth that we could not

---

* Each sub-species has different colouring.

climb up, but eventually we dislodged the corpse by moving it with three long bamboo poles tied together. While we were doing this I found an empty cartridge case.

When at last I was able to examine the nearly mummified body I saw that it was riddled with shot pellets. The father's head was tilted backwards, his arms were held out in a position which suggested that he had been clasping something and he was looking into space with a terribly alive expression.

Two days later when I found Coli and his mother near the tree where the father had been killed, they appeared heartbroken. Not long afterwards they moved into the safety of the forest; they were extremely nervous and very shy even of me, whom till recently they had trusted. But now whenever the mother saw me watching them she turned her head away and Coli buried himself deep in her fur. Neither of them ever played and apart from satisfying their need for food they just remained cuddled together, motionless, gazing into space. Their silent and dignified grief was more than I could sometimes bear. Once, while I watched them, I made a sketch – the first I had attempted since my accident two years before.

Although it would not bring the father back to life or to his family, I asked the Game Department and the Police to find the poacher who was responsible for killing the father and they succeeded.

A month later I heard the mother calling in the early morning and I wondered whether she had resigned herself to the death of her mate and was now trying to teach Coli how to roar.

Recently I was offered another colobus to replace the father. However, as Coli and his mother were so devoted to each other, I felt that this might disrupt their relationship. Furthermore, there is just a chance that in a year or two Coli may mate with his mother and raise another family within their present territory. Although man must do everything he can to preserve wild life from extermination, it can sometimes be dangerous to interfere too much, for animals are often able to find the best solutions to their own problems.

SKETCHES

113

Marsebit

Below:
Landscape
with
Giant Lobelia

119

122